Conceptual Foundations of Human Factors Measurement

HUMAN FACTORS AND ERGONOMICS

Gavriel Salvendy, Series Editor

Hendrick, H., and Kleiner, B. (Eds.): *Macroergonomics: Theory, Methods and Applications.*
Hollnagael, E. (Ed.): *Handbook of Cognitive Task Design.*
Jacko, J. A., and Sears, A. (Eds.): *The Human–Computer Interaction Handbook: Fundamentals, Evolving Technologies and Emerging Applications.*
Meister, D. (Au.): *Conceptual Foundations of Human Factors Measurement.*
Meister, D., and Enderwick, T. (Eds.): *Human Factors in System Design, Development, and Testing.*
Stanney, Kay M. (Ed.): *Handbook of Virtual Environments: Design, Implementation, and Applications.*
Stephanidis, C. (Ed.): *User Interfaces for All: Concepts, Methods, and Tools.*
Ye, Nong (Ed.): *The Handbook of Data Mining.*

Also in this Series

HCI 1999 Proceedings 2 Volume Set

- **Bullinger, H. –J., and Ziegler, J.** (Eds.): *Human–Computer Interaction: Ergonomics and User Interfaces.*
- **Bullinger, H. –J., and Ziegler, J.** (Eds.): *Human–Computer Interaction: Communication, Cooperation, and Application Design.*

HCI 2001 Proceedings 3 Volume Set

- **Smith, M. J., Salvendy, G., Harris, D., and Koubek, R. J.** (Eds.): *Usability Evaluation and Interface Design: Cognitive Engineering, Intelligent Agents andVirtual Reality.*
- **Smith, M. *J., and Salvendy, G.*** (Eds.): *Systems, Social and Internationalization Design Aspects of Human–Computer Interaction.*
- **Stephanidis, C.** (Ed.): *Universal Access in HCI: Towards an Information Society for All.*

HCI 2003 Proceedings 4 Volume Set

- **Jacko, J. A., and Stephanidis, C.** (Eds.): *Human–Computer Interaction: Theory and Practice (Part I).*
- **Stephanidis, C., and Jacko, J. A.** (Eds.): *Human–Computer Interaction: Theory and Practice (Part II).*
- **Harris, D., Duffy, V., Smith, M., and Stephanidis, C.** (Eds.): *Human–Centered Computing: Cognitive, Social and Ergonomic Aspects.*
- **Stephanidis, C.** (Ed.): *Universal Access in HCI: Inclusive Design in the Information Society.*

Conceptual Foundations of Human Factors Measurement

David Meister

2004

LAWRENCE ERLBAUM ASSOCIATES, PUBLISHERS
Mahwah, New Jersey London

Lawrence Erlbaum Associates, Inc., Publishers
10 Industrial Avenue
Mahwah, New Jersey 07430

Cover design by Sean Trane Sciarrone

Library of Congress Cataloging-in-Publication Data

Meister, David.
Conceptual Foundations of Human Factors Measurement / David Meister.
p. cm.
Includes bibliographical references and index.
ISBN 0-8058-4135-0 (alk. paper)
1. Human engineering. I. Title.

TA166.M395 2003
620.8'2—dc21

2003052863

Books published by Lawrence Erlbaum Associates are printed on acid-free paper,
and their bindings are chosen for strength and durability.

Printed in the United States of America
10 9 8 7 6 5 4 3 2 1

This book is dedicated to those HF professionals who believe in the continuing development of their discipline.

Contents

Preface

A preface serves as a means of justifying an author's writing of a book. It may also provide clues to the reader about what the author had in mind in writing the book.

This author admits to a feeling of dissatisfaction with the current status of Human Factors (HF) and its measurement processes. Like most people, HF professionals want simple answers to complex problems. They forget that there may be unexplored depths underlying the most common practices they perform.

Why explore these depths if everyone is happy with the status quo? A basic principle is involved: Unless the status quo is disturbed, there can be no progress in a discipline. Disciplines do progress and improve their effectiveness (however effectiveness is defined), but this does not occur automatically; there is no "invisible hand" guiding a discipline. Change must be orchestrated by those willing to question the status quo.

Measurement can be performed at two levels: (a) The most common is that of human or individual / team performance and (b) the least common is measurement of the performance of the discipline as a whole. Most of the following text discusses the first topic, but a few words need be said about the second.

If readers ask, as they should, why the effectiveness of a discipline should be measured, it is because (a) measurement to a large extent defines a discipline, and so measurement at the human level is inevitably tied in with measurement at the disciplinary level, and (b) measurement results may reveal ways in which the effectiveness of the discipline can be modified and its direction changed to make it more effective. In performing such an evaluation of HF, the following questions and parameters should be considered:

1. What significant changes have occurred in the discipline over time? In a scientific discipline like HF these changes are most often reflected in its measurement practices—the characteristics of its research. A comparative study was performed of the work published in the 1999 and 1974 Proceedings of the Human Factors and Ergonomics Society (HFES) (Newsletter, 2002a). The results suggest that not much has changed in HF research over the past 25 years.

2. To what extent do HF design specialists make use of research outputs in their work and to what extent do these outputs make a significant contribution to system design?

3. What capabilities has the discipline developed since its inception? Because HF is inextricably linked to technology, it is necessary to differentiate between external technological phenomena (like the development of computer capabilities) and internal phenomena, such as the way HF professionals respond to these external phenomena. Without question, the advent of computers made a great deal of difference to HF people, reflected in such advances as computerized devices for behavioral measurement, the development of human performance models, increased emphasis on cognitive functions, and a concentration of interest in information and information management. Behavioral measurement has become more sophisticated in the sense that computerized statistical analysis of data has enabled researchers to perform many operations faster than they could in the 1950s with Marchant calculators. On the other hand, the fundamental functions of measurement have remained much the same as they were in 1950 and even earlier. Still, the focus of our research is on the experiment and the questions studied are tailored to satisfy experimental design requirements. The research themes under investigation are much the same as they were in 1950, with the addition, of course, of interest in software and higher order cognitive functions.

4. One way of evaluating the effectiveness of a discipline is in terms of what it can do for its users. Looking at the capabilities of this discipline, it becomes evident that, although a great deal of data has been collected over 50 years of research, almost none has been combined and transformed into quantitative predictions of human-technology performance—although in the 1950s and 1960s there were earnest efforts to develop a predictive methodology (Meister, 1999). Despite computer searches of reports, the way data retrieved is utilized is very much as it was pre-World War II.

Theories of human performance are theories describing human responses to technology (i.e., psychological theories), but not those of how humans perform as part of complex system (which is what HF is concerned about). The basic psychological paradigm, stimulus-organism-response (SOR) still describes human behavior in HF tests, but must be vastly expanded to deal with HF and, particularly, with system parameters.

Measurement must also consider the beliefs and attitudes of HF personnel toward measurement in particular and their discipline in general. HF measurement cannot be reduced to a set of experimental designs; the really significant part of measurement is how the professional views and analyzes the measurement problem. The latter begins with the professional's selection of the questions to be studied and includes not only the development of study conclusions from resultant data but, just as important, their application to a wider world than the research environment. Any measurement text seeking to deal with real measurement problems must deal with these factors.

There are different ways of looking at HF measurement. The simplest viewpoint is to think of HF measurement in procedural terms: to accomplish a certain result and develop a set of conclusions, one does so and so. For example, to discover the effect of certain variables, perform an experiment and analyze the results statistically. Any number of textbooks tell exactly what to do in step-by-step fashion. Many professionals have this mind-set and do not even consider the possibility of alternatives.

HF professionals consider themselves scientists. Science requires asking the questions *how* and, particularly, *why*. A purely didactic, procedural discipline, festooned with lists of things to be done in a specified order is not science, but technique; those who lust for such order and procedure are technicians.

It is entirely possible for HF professionals to perform their daily work without asking themselves why they do what they do. It may be that most professionals are more comfortable in this mode. Some may, however, find intellectual excitement in asking these questions, which is why this book was written.

This book endeavors to examine the conceptual foundations of customary measurement practices and to stimulate the reader to ask questions about them. This will involve both the author and the reader in considerable speculation, because firm data on this topic do not exist. It may even require the empirical examination of the attitudes of HF professionals to many questions.

There may be some who see this book as only a series of essays, because its tone is not academic and didactic. It makes no attempt to tell readers what they should do, although there are many efforts to get them to think about what should be done.

Certain themes or assumptions dominate the book and the reader should be aware of these, even before reading the first chapter:

1. HF is a distinct, independent discipline, with its own needs, principles, and methods. These present certain problems that HF professionals must solve.

2. Measurement is the essence of a discipline; it defines and reflects that discipline.

3. The purposes that HF was designed to fulfill are *explanatory* (how and why did something involving the relation between technology and the human happen); and *implementational* (to aid in the development of human-machine systems).

4. HF, as a goal-directed discipline, is dynamic and moves from one state to another in an effort to satisfy the purposes of Item 3.

5. Successful application of research results to the development of nonresearch entities, primarily systems, is considered a validation of the usefulness of the discipline.

6. For progress to be made, it is necessary for every discipline to perform a continuing self-examination of its processes.

7. Covert factors underlie human performance and customary measurement practices. One reason for HF research is to make these factors overt.

8. The system and the operational environment are the fundamental referents of the discipline. This means that HF data and principles must relate to, explain, and assist human performance in the "real world." If they do not, they may be anything, but not HF.

9. HF as a discipline is influenced by sociocultural as well as scientific factors.

10. The major requirement producing tension in HF research and system development is the need to *transform* behavioral principles and data representing human performance into physical surrogates, that is, technological entities. Application requires the adaptation of behavioral knowledge to physical structures.

11. The conceptual structure (CS) of HF professionals defines and will, at least partially, determine the future of HF. Because the discipline depends on its professionals, their CS is as much a valid topic for research as any other in HF.

There is still much uncharted territory in HF to be explored. If this book has any value, it is to help advance that exploration.

David Meister

Chapter 1

Basic Premises and Principles

Because of the unique characteristics of the human factors (HF) discipline, its measurement also has special characteristics. To demonstrate this, a number of questions must be answered.

First, what is meant by the term *measurement*? In its broadest sense, measurement is the application of numerical values to physical and behavioral phenomena. For example, in its basic form, measurement involves being able to say that something has a quality "more than" or "less than" something else. Then attempts must be made to develop numerical values to represent "more" or "less."

This chapter does not go into greater detail about measurement in general, because this would involve rather abstruse philosophical concepts. This follows the tradition of most writers on measurement; they assume that the reader knows what measurement is and proceed immediately to talk about research or experimentation or experimental designs.

Second, why is measurement important to HF as a discipline and to HF professionals as individuals? The answer to this question is built into the philosophy of science. A discipline that cannot measure hardly deserves the designation, because nothing can be done with the discipline, that is, it cannot manipulate physically or symbolically the phenomena with which it purportedly deals.

Finally, what is unique about HF measurement as opposed to any other, and particularly psychological, measurement? Most of this chapter and the remainder of the book deals with this question. There are many excellent texts describing behavioral/psychological measurement: What makes HF deserving of special consideration? The topic of behavioral measurement is a well-established one. The only reason to write more about it is if something new and different can be said about the topic.

The following differs from the usual text on psychological measurement that emphasizes experimental design and its statistics. Although these are not ignored as one of the methods used in HF, the view of measurement here is a much expanded one, which takes account of the "pragmatics" inherent in HF.

WHAT IS SPECIAL ABOUT HF MEASUREMENT?

An answer to this question requires answering another question: What is so special about HF? If HF is special, then it is also possible that the measurement processes derived from it are as well.

HF is a behavioral discipline, a lineal descendant of psychology. The first HF professionals were experimental psychologists who were enlisted at the beginning of World Wars I and II to make behavioral inputs to the new wartime technology (see Meister, 1999, for a more extended historical description). It is also an engineering and system-oriented discipline, because the practical use of its measurement efforts is to assist engineering in system design.

Some professionals may think (see Hendrick, 1996) that HF is merely a variant of psychology. The fact that the behavioral input has to be applied and have an effect on a physical equipment means that a *transformation* is involved, thus breaching the barrier between the behavioral domain and the physical domain.

No other behavioral discipline (e.g., psychology, anthropology, sociology, etc.) involves such a transformation. Psychology does not concern itself with the design of equipment, nor with human–machine–system relations. Psychology includes a subdiscipline called industrial psychology, with certain superficial resemblances to HF, but industrial psychology is concerned only with the *effect* of equipment on the human and the equipment environment (the workplace, illumination, temperature, etc.) on operators; it is not concerned with the *design* of such equipment. Nor is engineering, which can be considered as the "stepfather" of HF, concerned with human operators and their behavior. This is true even though it has to take those operators into account in the design and operation of equipment.

HF is, therefore, something like biochemistry, a hybrid of two parent disciplines, but demonstrably a unique discipline of its own. Transformation, the human–equipment–system relation, and involvement with the design of machines are what make HF unique.

The uniqueness of HF makes its measurement processes also unique. This does not mean that there is no overlap between HF and psychological or engineering measurement; the output of measurement, human performance, is the same for all. Whether the human is a subject in a psychological laboratory or a worker in a process control plant, the human perceives, interprets, and responds to stimuli. So, the proverbial stimulus–organism–response (S–O–R) paradigm, which all first-year psychology students learn, applies also to HF, although in HF that superficially simple paradigm is tremendously more complex.

What makes the human in the work situation different from the human in a psychological laboratory is the *context* in which the person functions. What makes HF measurement different are the questions asked of it, the themes it pursues, its fundamental premises, and the environments in which it measures.

The aspects that characterize HF and psychological measurement appear to be the same, if attention is paid only to the mechanisms with which humans respond to stimuli. In both disciplines, stimuli can be presented only visually, aurally, or tactually; the human can respond only with arm or leg movements or conceptually (expressed in verbalisms). It is only when this same human behavior is viewed in the larger context of required task and system performance, with their individual purposes and goals, that the distinctiveness of HF performance and, hence, that of its measurement is revealed.

Because psychology is hardly concerned with machines, its attention is on humans alone and their perceptual, motor, and cognitive capabilities. There are other contexts in psychological measurement, such as learning, in which machines are found as instruments, but in psychology the machine as an essential element in human performance is largely lacking. The human–machine–system relation enforces—or should enforce—a choice of measurement topics that are foreign to psychology.

The human–machine–system relation is the fundamental element that distinguishes the HF discipline. Related aspects, like the task and the goal, are to be found occasionally in psychology, but are much less significant in that discipline. That is because tasks and goals are inherent elements of the system and cannot be understood except in a system context. Psychology does not deal with the system except as a source of stimuli.

Another differentiating factor is that the units of measurement in psychology are much more molecular (e.g., seconds and minutes) than they are in HF (e.g., hours and days). Tasks in HF are, because of their inherent equipment and mission contexts, likely to be much more complex than are tasks described in psychological terms.

Besides the human–machine relation, one other significant element in HF must be pointed out: Humans are an integral part of a larger system that organizes and, in part, determines their performance. The system, defined as a goal-directed organization whose human and machine elements interact (Van Gigch, 1974), controls the human performance whose measurement is a major concern of HF. This control relation is not found in psychology. In that discipline, tasks and goals are solely individual ones; tasks and goals in HF have no significance except in relation to the system.

WHY IS MEASUREMENT IMPORTANT?

The measurement processes involved in a discipline define and reflect that discipline. Some scientists assume that what cannot be measured does not really exist. (This is an extreme position, an exaggeration, of course, but only slightly so.) The definition of measurement used here may be more inclusive than that of most HF professionals. The analyses that produce hypotheses, speculations, and theory are considered essential parts of measurement; the conclusions derived from factual data are also assumed to be influenced by analysis.

Some readers may consider that measurement is an issue only for those performing research. Measurement is much more than a matter of research or the investigation of variables in predetermined ways. Measurement determines the characteristics of HF parameters (e.g., whether Task A is different from Task B), with which all professionals must deal.

This book is concerned with fundamental questions, such as the assumptions that determine measurement methods. In consequence, it does not provide simple rules of procedure; measurement, as the foundation of a discipline, is too complex for simplistic rules. Readers will almost certainly end this book with more questions than there are answers. However, questions must be asked in order to find answers.

ELEMENTS OF THE MEASUREMENT SYSTEM

The term *system,* which has so far been used to describe human–machine systems, can also be applied to other conceptual entities. Measurement is a conceptual system (as shown in Table 1.1 and Fig. 1.1), because it is an organization (arrangement of cognitive functions and physical elements made dynamic and energized by a goal, which is to solve a problem requiring measurement). Table 1.1 provides a structure for measurement as a whole, but only the first three elements of Table 1.1 are described in this chapter.

The measurement system consists of elements, personnel, processes, venues, outputs, factors, and attributes characterizing that measurement. Some of these are not part of measurement as such, but they do influence the process.

TABLE 1.1
Measurement System Components

Measurement Elements
1. Problems initiating measurement
2. Measurement assumptions
3. Measurement purposes and goals
4. Measurement questions and themes
5. Measurement criteria
6. Methods and types of measurement
7. Measurement outputs
8. Measurement applications and uses

Measurement Personnel
1. The individual who measures, including those who develop measurement devices (henceforth called the "specialist" or the "researcher")
2. The individual who uses the measurement output (the "user")
3. Measurement management (including funding personnel)

Measurement Processes
1. Selection of the measurement problem or topic
2. Search for additional information
3. Development of measurement procedures
4. Data collection
5. Data analysis
6. Measurement reporting

Measurement Venues
1. The operational environment (OE)
2. The laboratory
3. The simulator
4. The test facility

Measurement Outputs
1. Data
2. Conclusions
3. Theory/speculation

Measurement Influences
1. Nature of the human–machine system or the human function being measured
2. Available knowledge used as an information source (henceforth called the "knowledge system")
3. The operational environment
4. Availability of funding
5. The general intellectual atmosphere (henceforth called the "zeitgeist")

Measurement Attributes
1. Objectivity/subjectivity
2. Measurement validity
3. Data reliability
4. Generalizability of conclusions
5. Relevance of measurement outputs to operational reality
6. Utility

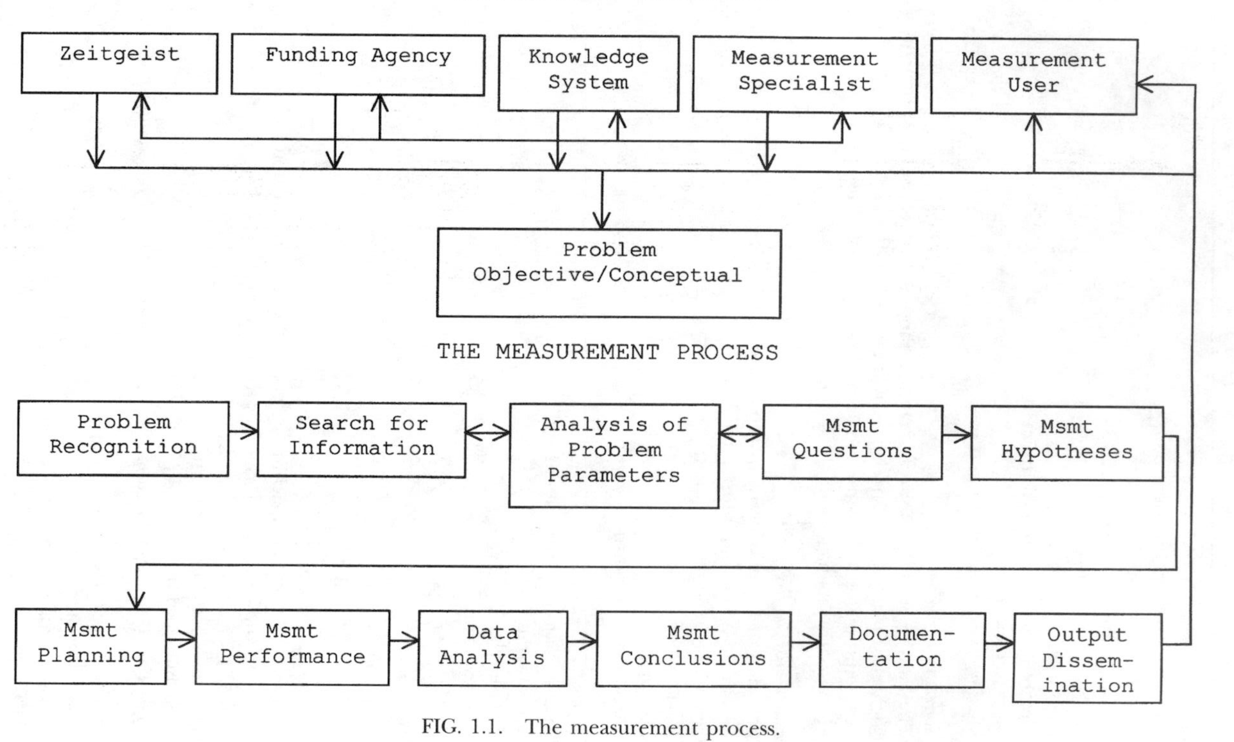

FIG. 1.1. The measurement process.

These elements are static as listed in the table. Figure 1.1 takes these elements and organizes them into individual processes to produce an overall process that moves (more or less sequentially) over time. Most of the interactions in Fig. 1.1 are feedforward; a few, like the search for information and analysis of problem parameters, also provide feedback.

The Measurement Problem

Measurement is initiated by the recognition that a problem exists requiring measurement. The problem may arise in either of two domains: in the real world of human–machine systems manifested primarily in system development (this form of measurement is called *system-related research*), and the conceptual world describing these systems and containing all available knowledge about those systems, its operating conditions, and its personnel (the measurement resulting is called *conceptual research*).

The real-world (objective) problem has been generated by a consideration of the status of the human–machine system. The system may be one under development or one that is operational, and the term *system* also encompasses its working environment (illumination, etc.) and the manner in which the system functions (its operating and maintenance procedures). The definition of measurement also includes the forensic situation, in which the human can be considered as a system unto itself. The discussion, however, centers on systems under development and in operational use.

The problem initiating measurement is a question relating to the status of the system. For example, does the system satisfy the design goal? Or, which of two or more potential system configurations is best? Or, in an operational system (e.g., a production system in which workers are experiencing an excessive number of injuries), where the problem is the question, what is the cause of these difficulties? Or a system may be redesigned to remove problems and the question is, has the redesigned system eliminated the problem?

Such real-world problems are an integral part of the system development process, because, as Meister and Enderwick (2001) pointed out, design/development is largely a search for information to answer the questions that arise during this process. Of course, not every system development problem/question is solved by the use of measurement. Indeed, formal measurement is fairly uncommon in system development compared with answers to questions derived by examining reports and documents, taking advice from colleagues (Allen, 1977) or simply accepting certain hypotheses as fact. In system development, measurement occurs only when the question cannot be answered in any other way. That is because measurement has costs that managers do not wish to incur, if there is some other way of answering the question. The kinds of system development ques-

TABLE 1.2
System Status Problems/Questions Requiring Measurement

1. Does the system under development perform to design requirements?
2. How well can system personnel operate/maintain the new system?
3. Have changes made to an operational system to eliminate problems been successful in solving these problems?
4. As between two or more system configurations, methods, procedures, training materials, etc., which is best?
5. Does the operational system, including its procedures and the environment in which it functions, perform effectively? What personnel-related design deficiencies still exist in the operational system?
6. What preferences do potential users have in relation to prototype designs?
7. What obscurities exist in a newly operational system?
8. How have personnel operated a predecessor system?

tions/problems that derive from the development process are listed in Table 1.2.

A single measurement (test) will often answer more than one of the questions in Table 1.2. Each of the categories in Table 1.2 is explained further.

Objective, Real-World Problems

The questions in Table 1.2 are behavior-related questions. They may be of some interest to engineer-designers, because all systems must be operated by personnel. They are, however, of greater interest to HF design specialists, because most engineers lack the background to understand their significance.

The following is not a detailed explanation of the categories in Table 1.2. Such an examination would require a much more complete discussion of human factors in design, which would be inappropriate in a measurement text (but see Meister & Enderwick, 2001).

1. Does the System (and More Specifically Its Personnel) Perform to Design Requirements? Through periodic testing during development, design personnel have a pretty good idea that the system will function as intended. However, until production begins, along with system verification testing, measurement has been performed not with the total equipment/system, but with modules, individual equipments, and subsystems. Even though the individual parts of the system are known to function to design specifications, it is not possible to be completely certain that the system as a whole will function properly, which is why Question 1 arises. For operators in particular, testing with parts of the system does not equal personnel exercise of the total system, because testing of components does not make the same demands on personnel as does the entire system.

2. How Well Can System Personnel Function in the New System? This question may be answered by the same test conducted to answer Question 1, but it is not the same inquiry as that put forth in Question 1. Question 1 asks merely whether personnel can perform (yes or no?); Question 2 asks, how well? Personnel may be able to operate the system, but only with difficulty, a high incidence of error, or severe stress. An answer that merely specifies that personnel can operate the system is therefore not sufficient. The degree of performance effectiveness must be determined to ensure that deficiencies impacting personnel performance do not exist.

3. Have Changes in System Design Eliminated Behavioral Problems? Particularly in a production subsystem (e.g., assembly line, paint shop, etc.), the subsystem as originally designed may produce behavioral difficulties such as delays, errors, and injuries. The cause of the difficulties must be ascertained (this, too, is part of the measurement process) and then, after the subsystem has been redesigned (e.g., the assembly line has been rearranged), performance in the redesigned subsystem must be measured and compared with its previous performance to see if the problem has been eliminated.

4. Which of Two or More Systems Are Best? During the course of development, two different equipments/system configurations, operating procedures or training materials, and so on, may have been proposed, but a decision between them cannot be made except by performance measurement. If the factor determining which of these will be selected is personnel performance with each alternative, then a comparison test will be performed. How often such a comparison will be made is unknown, because alternatives may be selected on the basis of factors other than that of personnel performance, such as cost or anticipated reliability.

5. Does the Operational System Perform Effectively? What Personnel-Related Design Deficiencies Still Exist in the Operational System? Even after the new system enters production, deficiencies may still exist. The system may then be exposed to special tests, as in the U.S. Navy's operational testing of new warships, or use by consumers to reveal such deficiencies. An example of the latter is beta-testing of new software (Sweetland, 1988), in which companies will give copies to selected users to use routinely over a period of months, thus providing an opportunity for users to discover and report any problems that may have escaped the attention of designers, and to explore use-scenarios beyond those initially envisaged by designers.

6. What Preferences Do Potential Users Have? Particularly for new consumer products, prototype testing of interim designs has been employed to discover user preferences (or aversions) in the new design, after which the design can be appropriately modified. Potential users are given the opportunity while the design is still under development to try out alternative versions and to indicate which they prefer and why. This is a means of allowing the user to participate in the design and to incorporate user desires in the final product. This measurement problem is not the same as that of Question 5. In Question 5, design has been completed; prototyping or usability testing is performed during design.

7. What Obscurities Still Exist in a New System? In very complex systems, the complexity may be so great that aspects of system functioning remain obscure, even after the system is turned over to the owner. Exercising the system in an operational environment enables the owner to explore these shadowy aspects and to compare subsequent modifications of the system (e.g., Version .01, Version .02, etc.). There are similarities here to Question 1 and, particularly, Question 4.

8. How Have Personnel Operated a Predecessor System? As part of the search for information during design, and particularly during its very early stage, HF designers may wish to learn how personnel have performed with a predecessor system, because much of what operators have done with the predecessor system may be carried over to the new one. Almost all systems have predecessors; from that standpoint, all design can be thought of as being essentially redesign of an earlier version. The specification of new human functions and tasks in the new system will be influenced by what has been done with earlier versions. Hence the need to discover "past history." The measurements involved in learning about past performance may be only partially performance oriented. They may involve observation of how the system personnel using the predecessor system go about their work (because in almost all cases the previous system is still functioning); often this is accompanied by interviews, surveys, and demonstrations.

Conceptual Problems

Conceptual problems, of most concern in the remainder of this chapter, arise from the examination/analysis of what can be called the *knowledge system*. In the broadest terms, the knowledge system is the totality of all that is written about, known, theorized, or speculated about a particular topic (e.g., cognitive task analysis; see Hoffman & Woods, 2000). It is a system because it is organized and the functions associated with knowledge usage are governed by the user's goal.

It is the researcher rather than the HF design specialist who is primarily concerned with conceptual problems, although in some cases the two may be the same individual. However, the data secured in the course of the conceptual research may be applied by the HF design specialist.

System-related research stems from real-world problems recognized by designers and specialists. Conceptual research stems from a conceptual problem (usually from examination of previous research and theory), but may occasionally be stimulated by an empirical problem. The types of problems that initiate conceptual research are listed in Table 1.3.

Although concepts drive conceptual research, it should not be assumed that conceptual researchers have much more freedom to select the topics they wish to study than do system researchers. The key intervening variable here is the funding agency, because almost no research is performed without being funded by some agency.

There are governmental and government-supported agencies tasked to study HF issues (e.g., the Office of Naval Research, the Army Research Institute), which employ researchers to study certain topics on their own and also let contracts to nonprofit research groups like the American Institutes for Research to study HF topics at the funding agency's behest. For conceptual problems, the choice of topic and measurement procedure is a matter of joint agreement between the researcher and the funding agency.

The conceptual problem that is ultimately studied is, therefore, subject to certain caveats. Researchers must be clever and experienced enough to find a suitable research topic. They must have a sufficient reputation as a researcher in the selected area to attract the attention of the funding agency; some funding agency must have sufficient interest in a conceptual problem to put up money to study the topic; and the topic must be important enough (in the eyes of the funding agency) to warrant its support.

TABLE 1.3
Types of Conceptual Problems

1. There is, in the opinion of the researcher, a lack of sufficient information about a problem area; this is by far the most common rationale presented in published articles.
2. The available information about a problem suggests to the researcher a hypothesis that should be tested.
3. Several published studies present contradictory information about the explanation of a particular phenomenon.
4. Measurement results of one or more tests involving a particular problem topic reveal ambiguities that must be resolved.
5. Knowledge-related products (e.g., a new method of assessing workload or eliciting information from personnel) have been developed; do these perform as specified?
6. Two or more knowledge products that serve the same purpose have been developed; which is more effective?

In consequence, there is nothing inevitable about the study of a conceptual problem. The general problem area may warrant measurement, but the potential funding agency may not like the potential researcher or what the researcher proposed to do in relation to that problem; it may not feel that the specific conceptual issue proposed for study is sufficiently important; or the agency may, at this moment, lack funding to support the effort.

Characteristics of Conceptual Problems

The following is based on a number of reviews of the HF literature (primarily of the Human Factors and Ergonomics Society [HFES] annual meeting proceedings) as summarized in Meister (1999) and Meister and Enderwick (2001).

The most common incentive for studying a conceptual problem is that not enough is known about a particular phenomenon or topic. It is part of the scientific ethos that there is always more to be learned. In consequence, the most common rationale for a research study is that not enough is known about a phenomenon or problem, and this particular study has been performed to remedy the situation. This rationale is, in large part, *pro forma*; if, in the introduction of a published article, it is necessary to explain why the study was performed, this excuse is as good as any.

Or it may be that a review of the available literature has suggested a hypothesis that should be tested. Ultimately, however, the underlying reason for the study is more mundane; most often, the funding agency was interested in the topic.

A conceptual problem may also arise if the results of a series of studies on a given topic reveals inconsistencies and contradictions. Study 1 says A, Study 2 says B, and Study 3 suggests that neither A nor B is entirely correct as an explanation of a phenomenon. The contradictions are inherent in the conclusions derived from the data. Presumably, no quarrel can be made with the data, but the interpretation of the data is something else again. Revisionism is rampant in science; the same data viewed by two researchers standing symbolically at different angles to a phenomenon often produce different interpretations.

It has also been hypothesized (Newsletter, 2002b) that nonexperimental "experience" in living and working with HF phenomena creates certain assumptions in professionals that partially determine the research themes they select and their hypotheses about variables. When experimental results become available, these assumptions also influence the professionals' willingness to accept or reject results that agree or disagree with their expectations.

In comparison with conceptual research measurement, system-related measurement is much simpler, because the latter does not ordinarily seek to discover how causal factors or invisible variables function. Conceptual research measurement is an order of magnitude more difficult, because the goal of research is to unearth these invisible variables. Because this cannot be done directly, researchers engage in a great deal of speculation and theory building that are nonetheless part of the measurement process. Speculation and theory are not merely a harmless amusement, because they can lead to testable hypotheses.

To try to understand the human–machine relation, researchers develop tools that must in their turn be measured. For example, workload is a construct that attempts to describe the relation between certain machine- or situation-induced stimuli and performance responses to these stimuli. A tremendous amount of research has been performed to develop test batteries to measure workload (Wierwille & Eggemeier, 1993). These tests, in turn, require measurement to assess their adequacy, to determine which are more effective than others. Thus, a conceptual problem (workload and its characteristics and effects) gives rise to conceptual products that, in turn, must be exposed to measurement.

In contrast to system development problems that have a finite and relatively short life because they are linked to a specific system, conceptual problems have an indefinite life, based on an indefinite interest.

The birth of a conceptual problem depends on the individual researcher, of course, but just as much on the zeitgeist, in which a major segment of the research community expresses an interest in a topic, as evidenced by the number of studies performed on that topic, its publication rate, the development of symposia on that topic, and the willingness of agencies to fund its research.

No conceptual problem is ever completely solved. System development problems are transitory; conceptual problems live forever, often in changed dress. For example, the memory problem with which Ebbinghaus struggled in the 1880s still exists as a problem, but in much changed form. The name of a problem (e.g., situation awareness instead of attention) and the orientation to the problem may, however, change. In consequence, there are an infinite number of problems existing in limbo (not yet born) or in an indeterminate life (in process of being studied, but not yet solved).

At first glance, there appears to be a sharp dichotomy between conceptual and system development problems. The former contribute to the knowledge system, whereas the latter, because of their specificity (in relation to particular types of systems) do not. Because of this, much more prestige is associated with conceptual research, and researchers presumably require greater skill; the problems they deal with are more complex, because they deal with the invisible entities called variables.

Belz (2001) argued that system-related work, in some aspects, is more demanding than conceptual research. As a systems engineer developing products, he needed solutions that are implementable. He did not have the luxury of merely interpreting findings and leaving the implementation to colleagues with the caveat that "more study is required."

THE CONCEPTUAL FOUNDATIONS OF HF MEASUREMENT

Assumptions and Principles

It is necessary to distinguish between measurement *assumptions* and measurement *principles.* The assumptions listed in Table 1.4 are *beliefs* derived from concepts of logic, science, and theory. Emphasis is on belief; these propositions need not be entirely or partially true, as long as they are commonly held by those who measure.

Measurement is conceived as a means of understanding and thereby controlling an ambiguous, uncertain world. This uncertainty, which is inherently repugnant to the human, leads to disorder. The function of measurement is to try to impose order through understanding. This may seem somewhat fanciful, but then why should anyone bother to measure anything? It is worth the bother to control that "anything" by understanding how it works. Everyday experience (not philosophy) suggests that considerable disorganization exists in the world. For example, computers are constantly breaking down. What can be done about this? Call a computer technician who "measures" the situation and restores the computer.

TABLE 1.4
Measurement Assumptions

1. The need to measure derives from a primary human need to understand an ambiguous, uncertain world.
2. Overt, visible behavior reflects the functioning of invisible mechanisms within the human.
3. The assumption of causation as regards human behavior suggests that behavior can be classified in terms of stimulus events, internal activity within the organism, and responses to the preceding.
4. Because all events appear to be uncertain, they can be understood only in terms of probabilities and predictions.
5. Measurement is always initiated by recognition of a discontinuity in the stream of events, which creates a problem, whether the problem reflects external real-world events or conceptual analysis of the state of knowledge.
6. Measurement produces diverse outputs (data, conclusions, and theory), only one of which (conclusions) has imposed on it scientific attributes (e.g., relevance, validity, reliability, generalizability, and precision) that are transformed into measurement goals.
7. Measurement requires and seeks control of the measurement environment.

The measurement principles described in the following section are less abstract than the assumptions, because the former are based more on the measurement experiences of individual professionals (hence are more concrete) than they are on logic and theory. To the individual specialist, they represent a strategy for solving concrete problems. Both assumptions and principles feed into and produce customary measurement practices (i.e., what every specialist does when performing measurements). The assumptions and principles accepted by professionals exercise an effect on measurement, the assumptions more remotely and generally, the principles more directly and specifically. The effect occurs even when the specialist is not thinking about these during the measurement process.

Basic Premises

The assumptions listed in Table 1.4 are extremely abstract and general, so at this level of discourse it is impossible to demonstrate the validity of the arguments presented, either empirically or through logic. Rather, they represent a point of view, one of several viewpoints perhaps, which can only be accepted or rejected, because at this level points of view have no inherent "truth."

The basic viewpoint expressed in Table 1.4 is that this is a world that is ambiguous and only partially understood. From this viewpoint stems the requirement to measure in order to seek further understanding and knowledge. A world in which people lack understanding is a potentially dangerous environment for the human. Measurement counters this danger and supplies a measure of confidence and satisfaction.

Table 1.4 also assumes that the act of measurement enhances control of this ambiguous world. Humans are assumed to possess an instinct to control their ecology (people, things, processes), an instinct that, on an intellectual level, is parallel to more basic needs for sustenance.

Explanation of Measurement Assumptions

The following explains the assumptions of Table 1.4 and is keyed to the listing in that table.

Assumption 1. The need to measure is one manifestation of the human need to make sense of a largely ambiguous, uncertain world. Or it may be thought of as a way to establish order (continuities out of discontinuities, certainties out of probabilities) in that world. Ambiguity produces real-world and conceptual problems that humans believe can be solved by measurement. The need to measure derives from the need to achieve certainty (or approximate certainty), because uncertainty is seen as dangerous. The lack of certainty, and a corresponding lack of or incorrect knowledge, is viewed as dangerous because it means that the human lacks control, and lack of control (over internal and external forces) can result in undesirable consequences.

Assumption 2. Overt, visible behavior merely reflects the functioning of covert, invisible mechanisms within the human. To understand and predict behavior, it is necessary to identify and discover how these mechanisms function. If humans are manipulated by forces not under their control, then it becomes necessary to discover what the forces are and to secure knowledge to help control them.

Assumption 3. All human behavior can be decomposed into the following: (a) stimulus events, (b) stimuli and responses occurring within the organism, and (c) responses to (a) and (b). This is the classic stimulus–organism–response (S–O–R) paradigm from psychology. The paradigm suggests a causal sequence, but immediate stimuli may or may not be the ultimate cause of a human response, because causal factors may be far removed from the stimuli they produce.

Assumption 4. All events are probabilistic or appear so, because, with limited foresight, it is not known if they will recur. This is true even for such common astronomical events like sunrise and sunset. There is a compulsion within the human for regularity, without which there is uncertainty. The only way to counter uncertainty is probability; it may be impossible to have complete certainty, but it is possible to approximate that certainty. To do so, however, requires predicting events, including an individual's own human performance and that of others. Prediction requires repeated measurement; it is possible to predict from a single occurrence, but the more data is available, the greater the precision of the prediction. Belz pointed out that experience is gained with practice; uncertainty varies with experience, and those with more experience are more certain.

Assumption 5. Uncertainty creates problems and one way of solving problems (other ways are by creating hypotheses or by establishing rigid rules) is by measurement. For HF there are two types of problems: real-world problems that stem from the development and operation of human–machine systems, and conceptual problems that derive from human beings' examination of their state of knowledge ("the knowledge system") and discovery of lacunae (they don't know enough), obscurities (what they know is confusing), and contradictions (measurements of multiple occurrences do not agree).

Assumption 6. Measurement has a number of outputs: data, which must be interpreted before they mean anything; conclusions, which are the result of data being interpreted; and speculations, hypotheses, and theories (which are the researchers' extension of the conclusion beyond what that conclusion tells them). Measurement also provides faith (a feeling of certainty), which may or may not be justified. Assuming the adequacy of the measurement process, data are simply what they are (i.e., bits of information). Data, when interpreted, lead to conclusions that have imposed on them by scientific logic certain attributes: relevance, validity, reliability, generalizability, objectivity, precision. All of these can be used as criteria to evaluate the conclusions and the knowledge system containing those conclusions. These attributes, then, become measurement goals, and goals help to determine measurement practices.

Assumption 7. To acquire knowledge, and to solve problems, the human requires control; a general principle is that what cannot be controlled becomes almost impossible to measure, although there are exceptions (i.e., planetary motion, tides, earthquakes). Control as exercised in measurement can be expressed in various ways (e.g., the selection of the topic to be studied; how that topic is studied; the selection of particular stimuli; selection of the subject's mode of response, etc.). Control also presupposes objectivity, because lack of objectivity leads to increased variability, and variability is simply another form of ambiguity.

Do Assumptions Have Effects?

It is necessary to consider whether the assumptions listed in Table 1.4 have concrete effects on measurement practices, because if they do not, they are essentially pointless. The following, again keyed to the list in Table 1.4, describes these effects:

1. The need to measure and to control has very direct effects. One effect is the overwhelming preference in the published HF literature for the experiment (see Meister, 1999, for empirical evidence of this). This preference derives from the fact that the experiment permits the greatest amount of control through manipulation of the conditions of measurement.

2. The concern for invisible mechanisms (variables) also leads directly to the experiment. The experiment is the single most effective method that permits isolation of variables through the categorization and manipulation of treatment conditions; these conditions presumably reflect the action of the variables. It is possible to isolate variables using other measurement methods (e.g., observation, the interview), but only with considerable difficulty.

The concern for variables also leads to the selection of certain research themes that center about the effort to isolate causal factors.

3. The S–O–R paradigm leads directly to the way in which measurement situations are set up: the selection of stimuli; control over the organism by selecting subjects with reference to some measurement criterion; and specification by the specialist of the way in which subject responses can be made.

4. Uncertainty, which results in probability as a counterconcept, thus leads to the requirement for prediction of human performance. All measurement leads to prediction, which is of three different types: *extrapolation* of the conclusion (If the same conditions that were controlled in the measurement are repeated, then the same set of conclusions will result from the data); predictions based on *combinations* of data (this has been called "human reliability"; see Meister, 1999. For example, if the average error rate in a number of studies is .0008, future measurements of the same human function will probably manifest an error rate varying around .0008); and predictions based on *theory* (e.g., certain mechanisms, variables, studied in one or more experiments will exercise an effect under conditions specified by the theory).

5. The implication of the preceding is that the nature of the problem will affect the nature of the measurement and the conclusions derived from the measurement.

6. The fact that conclusions are interpretive statements means that the interpretation can or will be influenced by biases or inadequate understanding.

7. The need for control in measurement leads again to the experiment as the primary method for ensuring control. It also leads to an insistence on objectivity in measurement, because subjectivity can lead to a lack of control.

Both assumptions and principles depend on belief, and belief depends on the individual. Although it is likely that some measurement beliefs are held more strongly than others, no one has investigated the extent to which strength of belief influences measurement practice. The more abstract a construct, the greater the influence of belief.

The fact that an assumption is widely held does not mean that it cannot be violated or ignored in actual measurement practice. For example, the belief that validation of conclusions is scientifically essential is often, if not usually, ignored in HF research. Does this mean that the validation assumption is not believed? No, but it does mean that the same individual who believes in validation can also believe that, in reality, validation is not necessary or that it may be unfeasible (a parallel and negative concept). It may be that some assumptions (e.g., validation) are inculcated through training, whereas others that negate these assumptions (e.g., validation is desirable but not critical) are inculcated through experience (reality-checking). Reality is likely to prove stronger than science. Belz (2001) suggested that for an applied discipline reality may be more appropriate.

It will, no doubt, have occurred to the reader that, if professionals are less than completely aware of the assumptions, principles, and beliefs that influence their measurements, then how could Table 1.4 and its implications be developed? This is an excellent question.

The answer lies in what is called "reverse logic," which is comparable to reverse engineering. If it is observed that certain measurement practices are common, then how could they have come about? For example, if the experiment is so much preferred by HF researchers, then what features of the experiment are so attractive? Then looking at the characteristics of the experiment (isolation and manipulation of measurement conditions, the search for invisible mechanisms-variables, and the great emphasis on control), it may be inferred from these that the researchers' belief structure *requires* these characteristics. From these, it is possible to work back and formalize these characteristics as conceptual requirements or assumptions.

Not all HF professionals will agree with these assumptions, because most are not discussed or even mentioned in standard textbooks (except for references to the common scientific attributes of validity, reliability, and objectivity). It is doubtful whether most HF professionals, even researchers who should be most concerned about their measurement concepts, think much or seriously about these concepts. If they do not believe that measurement is influenced by their assumptions, then they may not even consider the topic worthwhile.

Nevertheless, all professionals accept certain assumptions, whether or not they think about them. For example, they accept the scientific need for objectivity, and in consequence they denigrate subjective data (even as they use those data). They accept the need for valid conclusions, even though they may feel no personal need to validate their own work. They accept the need for control, and this leads them to espouse the experiment. Thus, at least a few of the assumptions in Table 1.4 (principally those inculcated by university training) are accepted by most professionals; and, presumably, those that are accepted exert some effect on measurement operations.

Measurement Principles

It is sometimes difficult to differentiate principles from assumptions. Assumptions are more abstract, more general than principles; principles reflect more directly customary practices. Assumptions subsume principles (which can, therefore, be deduced from assumptions), but professionals are more aware—if they are aware at all—of principles, because these are closer to what specialists actually do. As was the case with assumptions, principles are inferred from observation/documentation of customary practices.

Some, if not all, of the measurement principles listed in Table 1.5 have been anticipated by earlier material, because it is impossible to introduce the topic of HF measurement without referring to these principles. These principles include definitions and viewpoints, and their implications suggest a strategy for conducting measurement.

TABLE 1.5
Measurement Principles

1. HF measurement differs from other types of behavioral measurement, with respect to: (a) its focus on human–machine system relations; (b) the existence of real-world problems involving system development and operation that can be solved by measurement; (c) criteria of effectiveness or accomplishment, which are inherent in all HF problems; (d) the products of HF research measurement, which have real-world uses to which they must be put; (e) HF measurement, which can be evaluated on the basis of its utility. In consequence, measurement processes derived from other disciplines not possessing these characteristics have only limited applicability to HF.
2. All HF measurement stems from recognition of physical or conceptual problems whose solution requires measurement. Physical problems are produced by questions arising from system development and operation, and cannot be considered apart from design and operational processes. Conceptual problems result from analysis of the HF literature (the knowledge system) and lead to research that seeks to solve those problems.
3. Measurement includes analyses and other functions that precede and follow the physical acts of measurement; these involve speculation and theory, and include nonmeasurement functions such as documentation.
4. Measurement of systems involving personnel requires measurement of all relevant personnel performance involved in those systems.
5. Measurement functions and outputs are affected by contextual factors, including: (a) the amount and availability of knowledge relevant to the problem under investigation; (b) the biases of measurement personnel; (c) the nature of the problem to be solved by measurement; (d) the characteristics of the measurement object; (e) measurement purposes and goals; (f) measurement instruments, methods, and venues.
6. HF research has several purposes: (a) to discover the effects of selected technological variables on human performance; (b) to assist in the development, transformation, and inclusion of behavioral principles and data into human–machine systems; (c) to enable predictions of future human performance phenomena.
7. All HF measurement has an ultimate reference that supplies the criteria for evaluating the adequacy of that measurement. In the case of HF measurement, that reference is the operational environment (OE) or the domain in which human–machine systems ordinarily function. The most important of these criteria is relevance.
8. Despite the goal of maximum objectivity in measurement, HF measurement must take account of subjective factors, including subjectively derived data and judgmental values.
9. Measurement cannot be considered independent of the uses and applications to which measurement outputs will be put. These uses should be specified as part of measurement planning.

(Continued)

TABLE 1.5
(Continued)

10. All measurement can and must be evaluated in terms of the effectiveness not only of the individual study, but also and equally important, of the totality of measurement efforts directed at a general function and/or issues (e.g., research on decision making). Almost all measurement is directed at answering a specific question involving a specific device, population, or phenomenon. Each study can be subsumed under a higher order category (e.g., perception or decision making). It is therefore possible, and indeed necessary, to ask, when all the knowledge contained in these individual studies is combined, what is known *as a whole* about perception or decision making or stress (whatever the more molar category is). Criteria for evaluation of measurement effectiveness will ordinarily include relevance, validity, reliability, generalizability, applicability, objectivity, and utility. These effectiveness criteria apply as much or more to general knowledge categories as they do to the individual study. Indeed, the term *knowledge* is more meaningful in relation to general behavioral categories than to the individual study. Further discussion of this point is found in chapters 5 and 8.
11. All measurement is influenced by nontechnical factors, most importantly, the availability of funding, as well as (in the case of HF research) by the general intellectual environment or interest among professionals in specific problem areas (the zeitgeist).
12. Performance measurement in complex human–machine systems involving a mission must be performed at all relevant subsystem levels and mission phases.

Implications of the Measurement Principles

The following discussion is keyed to the principles listed in Table 1.5.

HF Measurement Is Distinctive. The most important implication of this principle is that the subject matter of HF research must be in some way descriptive of the human–machine–system relation. A measurement situation that fails to include this relation as a major variable and is focused solely on human variables is, whatever else it is, not HF research. Technology, its effect on personnel performance, and the way behavioral data and principles can be utilized in system design, must be major questions for HF research topics.

System development measurement already includes these implications, because the human–system relation is inherent in the system. Most professionals think of system development as "merely" the application of HF principles to design. They do not realize that design and system development are also basic research topics for HF, for two reasons: the importance of system design for technological civilization, and because design, whatever else it is, is a cognitive process.

The emphasis on system development in HF means that the outputs of research measurement must be capable of being applied to system development. It is not acceptable that research measurement should contribute only to the knowledge store, without at the same time assisting in the application of those outputs to the construction and operation of machines.

The extent to which these goals are accomplished is a measure of the *utility* of HF measurement and suggests that measurement of both system development and conceptual problems can be used as a criterion of the effectiveness of HF research.

The Problem as the Source of Measurement. The implication of this principle is that, at the stage of planning a study, it is necessary to examine the problem very deliberately and analytically. This is particularly important for research measurement because the common rationale for HF research ("more needs to be known about . . .") is insufficiently focused. The researcher should ask, "What specific problem am I solving?" If the problem cannot be delineated concretely, then the measurement loses much of its value.

Measurement Range. Measurement involves much more than the act of collecting data. The latter should be preceded and followed by intensive analyses to determine why the measurement is being performed; how well the measurement situation relates to real-world operations; what the anticipated data will be; assuming that such data are secured, what hypotheses can be developed concerning the meaning of the data; and the uses and application of the data. It is not acceptable to wait until after data are collected to raise these questions, because the actual data collected will influence their answers. The point is that the analytic aspects of the measurement are as important as the actual data collection.

System Measurement Must Include Human Performance Measurement. This principle may be obvious to the reader, but it is reiterated here because it may be necessary to argue the point with engineering managers who control system testing.

Factors Affecting Measurement. The analysis preceding measurement should include examination of the following questions: What knowledge is available that bears on the problem? Has the specialist made any assumptions about the research topic that have affected the way in which the measurement is being conducted? What is there about the nature of the problem and of the measurement object itself that can affect the way in which it is measured? Are the measurement purposes and goals entirely clear in the specialist's mind? Have the most efficient measurement instruments, methods, and venues been selected? The intent of this principle is to cause specialists to analyze their concepts as these relate to the specific measurement, in order to eliminate bias and rectify inadequacies.

Measurement Purposes. It is expected that the researcher will keep in mind the overall purposes of HF research and, in particular, that these will guide the selection of measurement topics. Special attention should be given to the study of system design as basic research. Anticipated measurement outputs should be examined in terms of their possible transformation into design guidelines and for quantitative prediction of human performance.

The Operational Environment (OE). The implications of this principle are that the OE for the system being measured and the environment in which it ordinarily performs must be examined to ensure that to the greatest extent possible the measurement situation includes those OE characteristics that could affect human performance. This principle also has implications for simulation; to satisfy this requirement and enable measurement to be conducted under controlled circumstances, the OE must be reproduced as part of the measurement situation. Measurement outputs must also be examined with regard to the solution of OE problems and the relevance of these outputs to the OE.

Subjectivity. Notwithstanding the requirement to make measurement as objective as possible, complete objectivity is impossible as long as a human is involved in the measurement. A complicating factor in implementing this mandate is: What is meant by objectivity, because there are degrees of objectivity, and what should be done about measurement methods and outputs that can be secured only through the human medium (e.g., interviews, scales, observations)?

Measurement Uses and Applications. This principle has been emphasized repeatedly in the previous pages; measurement without use is not acceptable. The main question is: What are these uses? There are *explanatory* uses (e.g., this phenomenon occurred because . . .) and *implementational* uses (this datum or conclusion can be applied to system development in the following ways . . .).

Explanatory knowledge is less satisfactory than implementational knowledge, because the former remains only theoretical (unverified), whereas when implementational knowledge is applied to a physical system, the application validates the knowledge. Perhaps this suggests an additional principle: Application of knowledge validates that knowledge as no other means of validation can.

Measurement Evaluation. Measurement itself as a process must be measured. The individual study must, of course, be evaluated in terms of its adequacy (e.g., appropriate experimental design, sufficient number of subjects), but individual study evaluation does not reveal much. All the studies published in the literature (those that describe a particular topic) must be evaluated to determine how much is known about that topic, and which questions/problems still remain to be answered. It cannot be assumed that if the studies are each individually adequate, all the research on a particular topic is also adequate. This is because the measurement questions inherent in a topic (e.g., workload) are more than the question each individual study addresses.

Nonmeasurement Factors. The contextual factor of concern is the zeitgeist, the intellectual environment, which is defined by the interest the research community has in a particular issue. That interest influences the availability of funding, so that influencing the zeitgeist (provided this can be done) can open up funding channels.

There are, of course enduring measurement questions, but these are often transcended by a heightened interest in special, "hot" topics. Research measurement money tends to flow to these. As long as measurement depends so heavily on money, there will be competition in securing funding. How can the researcher manipulate the zeitgeist to secure more money? This is not easy, but it has been done by (a) coining new and distinctive labels for phenomena, even when these are well known, thereby implying that something genuinely new has been found and should be studied; (b) pointing out the severity of real-world problems (e.g., accidents) to which the new interest area is related; and (c) prioritizing problems and putting the new topic at the top of the list.

Measurement at All System Levels and Phases. If the measurement object is a system with multiple subsystems and mission phases, human performance must be measured in all of these, depending, of course, on the relevance of that performance to the accomplishment of the system goal. (It is ordinarily unnecessary to measure the quality of the meals served by the system cafeteria.) The system concept suggests that all system elements are interrelated in some form of dependency. Because of this, performance at one system level and in one phase may explain performance at another level in another phase, including terminal performance.

The Distinction Between Data and Conclusions. Because of the orientation with which the researcher analyzes data, varying conclusions may be derived from the same data. The specialist has a responsibility to examine alternative explanations for the data before one conclusion is accepted to the exclusion of others. This is necessary, because conclusions can be affected by personal biases and the availability of relevant knowledge that might influence those conclusions. If researchers have linked themselves to a set of hypotheses, there is a natural tendency to view the data in terms of those conclusions. Conclusions must not go beyond the limits imposed by the variability, precision, and reliability of data.

How would these measurement implications be characterized? They represent changes in ways of thinking about measurement. However, the nuts and bolts of measurement do not change. They cannot easily change, because ways of measuring human performance are very limited. The more important changes in viewpoint are those that result in changes in the selection of topics to be studied, the range of measurement interests, and the way in which data are interpreted. Such changes are more important than, for example, developing methodological devices such as a new performance model or a new questionnaire.

ACKNOWLEDGMENT

The author is indebted to Stephen M. Belz for a critical reading of this chapter and his suggestions for revision.

Measurement Problems

Tables 2.1, 2.2, and 2.3 list the functions ordinarily performed in the course of a measurement study. Every function in the tables is essential, but the most important ones have been indicated by an asterisk. Those so indicated involve uncertainty, because they require the researcher to make choices and decisions.

MEASUREMENT PHASES AND FUNCTIONS

Planning the Measurement

Selection of a research theme depends on the recognition that a problem in a research area exists. Each phase in Tables 2.1, 2.2, and 2.3 contains one or more problems. However, most of these problems, which are primarily conceptual, arise in the Planning and Data Analysis and Documentation phases. Operations involved in actual data collection are more or less "cut and dried," because the conceptual problems are assumed to have been solved in the planning phase.

The problems arise because the requirements for performing the phases are not easily accomplished. These problems, which affect the efficiency with which professionals conduct their research, are not merely theoretical; they have very specific effects that can hinder the measurement and affect the accomplishment of HF goals. For example, if measurement results cannot be readily applied, a major purpose of HF—to aid system de-

TABLE 2.1
Planning the Measurement Process

Preplanning

*1. Selecting the research theme.
*2. Recognizing that a specific problem requiring measurement exists.
*3. Determining the parameters of the problem.
*4. Determining the questions that must be answered by the study and the variables involved.
5. Determining what information is needed to pursue the study.
    (a) Determining where the information can be accessed.
6. Searching for the required information.
    (a) Determining what information is available.
    (b) Determining the measurement implications of the information.

Planning

*7. Development of initial study hypotheses.
*8. Determining which measurement methods (experimental, nonexperimental) are suitable.
    (a) Analysis of which method is most effective.
    (b) Selection of the performances to be measured.
    (c) Selection of the most effective measurement method.
9. Selection/development of the measurement instrument.
10. Determination of the type and number of subjects.
11. Determination of the most appropriate measurement venue.
12. Determination of measurement procedures to be followed.
*13. Selection of the measures to be employed.
14. Development of hypotheses about what the data will reveal.
15. Writing a study plan (optional).

*Refers to high uncertainty functions.

TABLE 2.2
Pretesting and Data Collection

Pretesting the Measurement Procedure

1. Pretesting the measurement procedure.
    (a) Collecting sample data.
    (b) Analyzing the sample data.
    (c) Determining that the measurement procedure is adequate or requires modifications.
    (d) Modifying the measurement procedure, if required.

Data Collection

2. Collection of measurement data.
    (a) Briefing/training data collectors.
    (b) Performing data collection.
    (c) Debriefing test subjects.
    (d) Closing down the measurement situation.

TABLE 2.3
Data Analysis and Documentation

Data Analysis

1. Data analysis
   *(a) Determination that the data answer the study problems/questions.
   (b) If data are not adequate to answer the measurement questions, then determine what to do next.
   *(c) Determining the conclusions suggested by the data:
      1. What system development fixes are required?
      2. What operational changes are required?
      3. If the measurement is a research study, then what applications of the data can be made?

Documentation

2. Writing the research report
   *(a) Determination of the measurement implications; how much of the initial problem has been solved?
   (b) Determination of what further information (if any) is needed.
   (c) Evaluation of the efficiency of the study; what lessons can be learned about measurement processes from the study?

velopment—cannot be accomplished. Implicit also in these measurement phases are questions about how researchers accomplish their functions.

Some measurement problems are not serious enough to hazard the measurement. In any event, researchers tend to ignore problems they feel they can do nothing about. Indeed, this attitude of helplessness before major problems like validation of research results, for example, tends to enhance an attitude of passivity in researchers. Part of the reason for this passivity is the assumption that every measurement has some undefined amount of error associated with it. The use of probability statistics strengthens the assumption. In any event, as long as researchers do not know the nature and extent of the error, they are presumably within their rights to disregard it (see the later discussion on "pragmatic assumptions" in this regard).

These problems are not the result of specific errors on the part of researchers, but are inherent in the intellectual challenges engrained in the measurement process.

Preplanning the Measurement

Selecting the Research Theme. Selection of the research topic is arguably the first and most important problem the researcher encounters. It is a reasonable assumption that no measurement is made unless the researcher and those who fund the research recognize that a problem exists that requires the measurement. The problem may be a purely empirical one. If, for example, an agency is attempting to encourage the use of technology by the elderly, then a study of how much technology is actually used by seniors, their attitude to technology, and so on, would seem to be warranted.

Other problems are conceptual. If the agency wishes to discover the causal factors for the presumed lack of interest by seniors in technology, then its researchers might create an hypothesis or even a theory to explain the lack of interest and would then perform a study to determine if the theory was correct.

It might be thought that selection of a research theme is solely the responsibility of the individual researcher. However, research must be funded, and funding agencies decide on their own what they will support. Research topic selection is still, however, a problem for the researcher, because all the agency does is to select a general area, within which specific topics must be developed. The latter is often the responsibility of the individual researcher, because many agencies do not concern themselves with research details until a proposal is made to them.

For example, assume that the agency's agenda is aging. Within this very general category the agency decomposes aging into a number of topics, one of which is technological impact on the elderly. Technological impact may be further categorized into subtopics such as social/cultural, learning, or financial. These categories are advertised and researchers interested in any of these topics are invited to submit proposals. The researcher must then select one of these categories into which to slot a specific research proposal. If, for example, researchers select learning as a subcategory, they might propose investigating the difficulties of training the elderly to accomplish new technological functions, such as computer operation. (A whole line of research has emphasized this interaction of technology and the elderly; Czaja, 1996, 1997; Rogers, 1999.) It then becomes the agency's responsibility to decide whether it wishes to fund a study of computer training of the elderly.

On the other hand, researchers may take the initiative; if they wish to study computer training of the elderly, they will look about for an agency interested in this topic and submit a proposal. If the proposal fits the agency's agenda, if the proposal is sufficiently innovative (others have studied this topic also), if the researcher's credentials are sufficiently impressive, and if the cost is not exorbitant, then the agency will issue a grant or contract.

Determining Parameters, Variables, and Questions. As part of the development of a specific research theme it is necessary to specify the problem *parameters,* the *variables* involved in the problem, and the specific *questions* that the study will ask. One has to do the same thing in system status problems, but the answers are much easier to secure with these last, because the nature of the problem (e.g., Does the system perform to design specifications?) provides all of these things.

In the selection of a research theme for a conceptual problem, consider this sequence: determination of the research topic (e.g., menus), followed by determination of the problem parameters (e.g., menu length and breadth), followed by the question the study will answer (e.g., Does the length of the menu affect the operator's performance?).

Another example: The researcher is already aware of the specialty area (e.g., individual differences) and now has to focus on the subspecialty area (e.g., fatigue), the parameters involved (e.g., length of time on the job, conditions of performance, etc.), and the question the study will ask: What is the effect of fatigue on decision making?

With regard to the variables to be manipulated in a particular experimental study, it is likely that the researcher is already aware of these, before study planning even begins. The information collected by researchers as part of the continuing search for research information will identify these variables.

The questions the study will attempt to answer and the hypotheses that are developed in advance to answer these questions are not the same. Questions come first, hypotheses next. Take research on the relation between alcohol and the aged. One of the questions the study may ask is: What is the extent of the alcoholism problem in the aged? The following may be related hypotheses: (a) Alcoholism is significant only in a subset of the elderly (i.e., those with only minimal financial and social support, little education, few relatives). (b) The effect of alcoholism is to cut off certain avenues of social help that are available to the elderly.

Parameters in the alcohol research problem are age, physical health, mental health, work status, family status, financial status, and so forth. These can also be considered variables, and hypotheses can be developed about each of them (how these variables function to influence the elderly).

In all that has been said, it is necessary to distinguish between system-related and conceptual researchers. For the former, the selection of the research theme is predetermined by the special problems faced by a specific system or systems in general. The number of study topics is limited by the central questions involving systems: How well does the system perform? Which configuration is more effective? How well do personnel perform? Has a redesign accomplished its goal? Everything in system research is much more straightforward than in conceptual research, including data analysis and application of results. Even the problem of validation (which is discussed later) does not really exist for system research, because the study answers are designed to relate to a specific system being tested.

There is, however, one type of system-related research that is akin to conceptual research. That research attempts to answer the following question: What factors in the architecture of the system as a whole (e.g., complexity) influence operator performance, and how does the effect occur? Unfortunately, few studies reported in the general literature are of this nature (except perhaps those dealing with software; examples are to be found in the *International Journal of Man–Machine Studies*).

For the other types of system-related studies whose questions were listed earlier, a general principle related to a system component like menus, for example, may be derived from the study results, but, if so, this is serendipity.

Unless the research is very novel or the subject's task is entirely covert, the parameters of the activity being investigated are probably reasonably evident. Previous research will have examined these, which is why acquaintance with previous research is so necessary. The researcher's personal experience with the study tasks will also be helpful. In system studies, parameters have usually been extensively reviewed in performing an earlier task analysis, or, if not, can at least be observed in performance of a predecessor or related system. Parameters may be more or less important, depending on the nature of the activity being investigated. For example, in research on the elderly, age, gender, ethnicity, finances, and health are clearly important; in a radar detection study, the operator's skill, the nature of the target, and its movement or nonmovement are all-important, but not ethnicity.

Parameters may also become more or less important, depending on the questions asked by the study. If the study deals, for example, with frequency of driving by the elderly, age, sex, health, and finances may be more important than, say, educational background, which last might become more important if it were an investigation of social relationships.

Parameters and questions give rise to hypotheses, but not automatically. The researcher has to imagine the potential functioning of the parameters. Parameters become variables as soon as how the parameter might influence performance is considered. For example, in considering the relation between age and use of technology, other parameters such as extent of previous technological experience, gender, and educational background might be mediating (interacting) factors. Older women may have had less experience with technology as represented by computers, and those with more advanced education may be more likely to have been exposed to computational and other devices as part of their education.

Parameters and questions are therefore organized by the researcher into a set of testable hypotheses. These form the basis for the development of treatment conditions in an experiment. In a nonexperimental study, the hypotheses serve to determine which subject activities will be observed and the way they will be measured.

The researcher is also motivated to develop hypotheses by the effort to discover causal factors. The hypotheses represent those factors.

In developing hypotheses that will direct the study, the researcher is betting that the hypotheses selected have a reasonable probability of being verified by the measurement. It is not logical to develop a hypothesis that is deliberately weak or inherently unlikely to be verified by the study. Put into probability terms, the researcher is betting the odds to win. An example of an unlikely hypothesis is that elderly driving performance is partially determined by the "make" and color of the automobile being driven.

Developing a Test Plan. Assume that most researchers develop a research plan along the lines of Tables 2.1, 2.2, and 2.3. Not only funding requires a plan, but if the research is part of an academic program, approval must be received from higher authority. Even if researchers are in the unlikely situation in which they are completely free to study anything and need not rely on an eternal agency to supply funding or to approve the research topic, clever researchers will develop a test plan as a checklist to ensure that all necessary steps are taken.

Collection of Relevant Information. As part of the preliminaries leading to development of a test plan, an effort must be made to compile all information from previous research results and to observe any research-related phenomena. Collection of such data is likely to precede or at least to be in parallel with development of a test plan. Obviously, there are problems in developing such information; all information sources may not be available, and the information may be incomplete or contradictory.

Planning the Measurement Process

Selecting the Measurement Method. A decision must also be made about the measurement method to be employed. The general methods are quite few: nonexperimental and experimental. Within these two broad categories (which are discussed in much greater detail in later chapters), researchers have a choice of objective and subjective methods or both.

HF researchers, like most scientists, have a definite preference for the experiment and for objective methods. The preference applies to both system and conceptual research. Only one type of system-related study requires an experiment; when the question to be answered is which of two or more configurations is more effective, a comparison strongly suggests an experimental method. When the question concerns how well the system performs, no experiment is required, even though or because the system is performing against some standard or criterion that represents the design-required performance. That standard can serve as an implicit treatment condition.

The choice of an experiment over a nonexperimental method is determined by a number of factors, one of which is that use of an experiment permits the researcher to make a more clear-cut comparison between alternatives. Particularly where causal factors are of interest, the hypothesis becomes that Factor X is responsible for a particular performance; this is contrasted with an alternative and correlated, although unarticulated, hypothesis, in which Factor X is not responsible for a particular performance or performance difference. Setting up the causal hypothesis into an *is* and *is not* configuration makes it relatively easy to develop treatment conditions to test the proposition. The experiment does not create the causal Factor X hypothesis. This is performed by the researcher as a prerequisite to the development of treatment conditions. The *is* and *is not* hypotheses may be responsible for the development of the basic method of contrasting the experimental (is) with a control (is not) condition. All more elaborate experimental treatments are simply expansions of the basic comparison.

The use of the experiment thus permits a clearer comparison of the is–is not hypotheses. There are additional factors involved in selecting the experiment as the measurement mode. Primary is that the experiment permits researchers greater control over what their subjects will be doing during the study. The treatment conditions place subjects into two or more mutually exclusive categories: those who receive the *is* condition (the experimental group) and those who do not (the control group). (There are other statistical designs in which, e.g., subjects receive all stimuli but in a staggered order to cancel out undesirable learning effects. These are, however, simply variations of the is–is not logic. For descriptions of statistical methods for setting up and analyzing treatment conditions, see Box, W. G. Hunter, & J. S. Hunter, 1978, and Winer, Brown, & Michels, 1991).

Control is exercised in the experiment because the treatment condition strips away all extraneous and possibly confounding variables. Factors that might affect the *is* condition are eliminated.

Other factors that influence the predisposition to the experiment are a variety of social and cultural influences. Among physical scientists, the concept of the experiment has long been considered the *ne plus ultra* of science. This attitude has permeated the university and been transmitted to its students, including, of course, budding HF scientists. Publication, which is the scientific Holy Grail, is difficult to ensure unless the study presented for publication is built around an experiment (including also its statistical analyses). One criticism that might be levied against HF research is that it may substitute statistical analysis for logical thinking.

Certain charges can be raised against the experiment, notably that it is artificial (who ever heard of operational performance being conducted like an experiment?). If the ultimate referent of HF research is the operational system functioning in its operational environment, then the ability to generalize research conclusions to operational systems is obviously limited by the experiment's artificiality.

As a contrast to the experimental situation there is the nonexperimental methodology. It might be assumed that this methodology does not allow the researcher to contrast conditions and therefore cannot be used to investigate causal factors, but this would be a misstatement. Alternative conditions may be inherent in the measurement situation; these can be contrasted even if the usual treatment conditions cannot be manipulated by the researcher. Conditions like different mission phases, different times of the day or contrasting weather conditions, and so on, may be inherent in the way a particular system functions normally. These can be contrasted to determine if these conditions are responsible for varying human performance.

If experimentation can be viewed as the most desirable measurement methodology, then under what circumstances is experimentation impossible or less desirable than its alternative?

The essence of experimentation is that it permits subjects to be moved around to fit into boxes representing the *is* and the *is not* hypotheses. There are situations in which subject manipulation is difficult or impossible. This is often the case in system-related studies where the total system is the vehicle for the measurement. Take a ship, for example. In its ordinary functioning, the ship is operated in only one way. If researchers wished to study the effects of causal variables on ship personnel performance, then it might be necessary to divert ship personnel from their normal tasks, as a result of which some hazard might result. The managers of the system (the plant manager, the pilot of an aircraft) may not permit any manipulation of their systems in order to establish contrasting conditions. The researchers may have to accept a nonexperimental methodology that consists of the recording of personnel performance as they go about their normal business. As a result, the factors responsible for performance have to be inferred rather than tested. That is, the researchers must look for *clues* to the factors responsible for varied performance. The nonexperimental researcher will undoubtedly have developed hypotheses about causal factors, but these may not be testable, because only inferences are permitted.

Many variables, such as purely idiosyncratic variables like intelligence, skill, or visual acuity, are inherent in system operating procedures, because these variables influence individual task performance. However, these are not important unless they directly influence overall system performance. Other variables are a function of the type of system and how it is operated (e.g., adversary threats in military systems), and these are of interest and can even be experimentally compared if a simulator or a human performance model is used. Even if these variables cannot be experimentally manipulated, their occurrence in routine system operations can be recorded and statistically compared.

Another decision has to be made between objective and subjective (i.e., self-report) methods of gathering data. If an experiment has been selected, then the choice will almost always be in favor of objective data. The choice of an experimental methodology, however, does not mean that subjective data cannot be collected, but this will ordinarily be feasible only following the collection of the experimental, objective data.

There will, however, be situations in which objective data cannot be collected. If the subjects' performance is entirely covert, such as their opinion or perception of how an event proceeded, then researchers can only ask the subjects about this. Unless someone has already developed a standardized subjective measurement instrument like a questionnaire or scale such as TLX for workload (see Gawron, 2000, for its description), researchers will have to develop such subjective instruments on their own. This is particularly the case in debriefings, when the questions to be asked are specific to the previous objective test. Many researchers may not wish to incur this effort, but it is necessary to study subjectively expressed variables.

Even if researchers are employing objective methods (e.g., instrumentation to collect data), they may wish to secure information from the subject that only the subject can provide. A study conducted by the author revealed that 24% of all papers published in the 1999 *Proceedings* of the HFES annual meeting utilized some form of subjective data. The researchers may feel that this information is ancillary and therefore not absolutely required, but often the test participant can provide interesting and perhaps useful information, and should therefore be debriefed following the test. For example, if a task performed by subjects is complex and, hence, especially difficult, then it may be useful to discover if subjects understood what they were doing.

The experimenter may feel that the constraints imposed by treatment conditions effectively control the influence of individual differences, but the fact is that these individual differences determine whether treatment conditions will be statistically significant. Hence, it would be useful to discover some of the factors producing these individual differences.

Selecting Performances to Be Measured. Not every function/task performed in system operations is equally important in achieving system success, and with limited resources the researcher may wish to concentrate on some (e.g., communication), while excluding others (e.g., clerical tasks) from the measurement process. Some system tasks are operating tasks, whereas others involve maintenance, and the researcher may choose to emphasize one or the other, or both with differential emphasis. Certain operational tasks like command decisions may be of primary interest and others, such as communication or detection, may be of lesser concern.

The selection of functions/tasks to be measured is important, not only in system-related studies, but also in conceptual research. What one presents to a subject in an experiment does not represent a random choice. The stimulus presented requires a task to be performed, and because the various ways in which humans can respond are limited, the task developed for an experiment is likely to be one found in actual systems or one that closely resembles such a task. The only things the human can be asked to perform are to detect, recognize, analyze, interpret, manipulate, track, and so on, the stimuli presented; all of which functions are to be found in system operations. The questions asked by a study will also determine the kind of function/task to be performed by the subject. The important thing for the researcher is to anticipate the effects the stimuli and task impose on the subject. The more difficult these are, the greater the likelihood of higher error frequencies and errors may increase subject variability.

In system-related studies, the nature of the stimuli and the task are determined by system requirements. But, in conceptual research, experimenters have more control and flexibility over what they decide to present.

The measurement situation as a whole requires examination. Among the elements of that situation are the following:

1. The functions/tasks already discussed, with particular attention to response time permitted; any provision for performance feedback or assistance to subjects; the opportunity for subjects to consult with other subjects (i.e., team vs. individual performance), permitted alternate response strategies, and so forth.
2. The nature of the instructions given to the subjects as to how they must respond.
3. The total time allotted to subjects to perform.
4. The measurement venue (where the study will be conducted).
5. The flexibility (if any) permitted to subjects to select the stimuli they will respond to and the nature of the responses they can make.
6. Any contingency relations of stimuli (i.e., whether individual stimuli are related to each other) so that the subject's performance with preceding stimuli will at least partially determine which subsequent stimuli will be presented.

Some of the previous situation elements are meaningful only if the conceptual researcher intends to replicate in a controlled format the characteristics of operational stimuli. The more the research task simulates an operational task, the less freedom the investigator will have to manipulate that task. On the other hand, the more the similarity between research and operational tasks, the more generalizability of the research results.

Selecting the Measurement Venue. Something should be said about the venue (laboratory, simulator, test site, operational environment) in which the measurement will be made. Factors of control and artificiality are important here. The laboratory permits the greatest amount of control, which is important for experimentation. The operational environment (OE) permits very little control, if control is defined as the opportunity to manipulate conditions of presentation.

Artificiality is defined by the lack of similarity of the measurement situation to the operational system functioning in the OE. For purposes of validity and generalizability, the researcher's wish would be to have maximum identity between the measurement situation and the OE. The experiment, with its manipulation of subjects and treatment conditions, is maximally dissimilar to the operational system and the OE; this is a primary deficiency of the experiment. The simulator offers similarity to the OE and control, but only a few simulators are available and they are quite expensive in money and time to develop. The test site requires a functioning system, but also provides more control than the OE. The OE provides maximum realism and least control.

Nontechnical factors, primarily money, may determine the venue for researchers; they may have no choice. On the other hand, there are alternatives. For example, a driving simulator may be unavailable, but it may be possible to make do with a driving track (a test site) or even measure performance on city streets.

Selection of Subjects. Similar considerations apply to the selection of subjects. If a specific subject population is required, like aircraft pilots, senior citizens, or hospital patients, then the researcher will have to make special accommodations for them. This is particularly the case when the nature of the subjects (e.g., paraplegics) restricts their capability to respond to stimuli.

If the nature of the system or the questions asked by the study do not require special subjects, then subject selection is much less onerous. Subject selection also depends on the nature of the population to which the research results are intended to be generalized. A general population is easiest to accommodate. Individuals who are human are by definition qualified to act as subjects, providing they can respond (as, someone who could not speak English might be disqualified). Nevertheless, considering the mischief created in statistical analysis by individual subject differences, general subjects may be too much of a good thing, because they may have unknown tendencies that could bias the data. There is always the possibility that for a particular study a group of general subjects may conceal factors that will adversely affect their performance. It would seem logical, then, for researchers to examine the subject population closely, but most researchers do not do this.

As for the number of subjects required, the researcher's main preoccupation is to secure as many subjects as possible within the constraints of test time, money for subject fees, and so forth. Formulae exist to determine the number of subjects needed to apply significance of difference statistics (see Gawron, 2000; Nielsen, 1997; Virzi, 1992), but although a minimum number can be specified (in most experiments, at least 20), the maximum number can be whatever the researcher can secure. For significance of difference statistics, the more subjects there are, the more likely it is that treatment differences will cancel out individual subject differences.

Selection of Measures. Another critical decision point in planning the study is the determination/selection of measures. First, of course, researchers must decide on how the subject will be asked to respond, but even after making the decision, the specific measures to be used must be selected. These are partially determined by the tasks that the subjects will be asked to perform. For example, if the task involves detection of, say, an enemy task force at sea, measures may be duration until first detection, duration until target identification, and so on. None of the elements of the measurement situation in Table 2.1 are independent of each other.

A detailed discussion of measures is taken up in chapter 7.

Pretesting and Data Collection

It is always wise to pretest data collection procedures, because in an uncertain world all sorts of problems that could only be hazily conceptualized in advance may arise. Pretesting enables researchers to discover these problems and make the necessary adjustments to eliminate them. An additional function of pretesting is to give researchers confidence.

Once pretesting provides this confidence, the actual collection of data becomes more or less routine, because unanticipated events have been anticipated.

Data collection itself is tedious, repetitive, and boring, but should pose no significant problems if measurement procedures have been pretested. The only anxiety producing element in data collection is the researchers' anticipation of the data coming in and their very preliminary analysis of those data to see if initial hypotheses are being verified.

Data Analysis

This chapter does not attempt to describe details of statistical analysis. The topic has been described in many books with a degree of detail that could not be matched here.

From a pragmatic standpoint, the major question the data analysis in conceptual research seeks to answer is: Have the initial hypotheses been verified with at least minimal statistical values—which in most cases is the traditional .05 level of probability or less? To most investigators, if this significance of difference level is achieved, then the researcher is "home free."

Beyond that, the statistics (e.g., the analysis of variance, ANOVA) must be scrutinized in detail for the interactions of the primary and secondary variables. If the initial hypotheses have not been verified at the .05 level, then it is necessary to discover why. Once the initial hypotheses have been verified statistically, the more demanding analysis begins: the translation from statistical data to qualitative conclusions.

There are two types of data analysis: statistical and qualitative. Both are often performed concurrently. The concern here is with the analysis that provides *meaning* to the statistical statement. The statistical analysis answers only the question: Do the data support the original hypotheses? The more detailed data analysis deals with how far beyond the statistical statement it is possible to go in explaining the study's conclusions.

If the original hypotheses are not supported by the data, then the researcher has the problem of damage control: What went wrong and what do the data, whatever they are, tell the researcher about what actually happened in subject performance? It may even be possible to restate the original hypotheses so that the data support the revised hypotheses. It is possible also that the revisions have something worthwhile to tell the reader of the research report. In any event, most hypotheses in published papers are supported by the data (or the paper would not be published). The researcher makes an effort to ensure that only likely hypotheses are studied.

Determination of Conclusions. At this point, it is necessary to derive conclusions that are developed immediately prior to writing the study report. The term *conclusions* does not mean merely the verbal restatement of the statistical analyses. Researchers wish to show the larger significance of what they have found. To do so, they must go beyond a simple restatement.

Suppose, for example, a study was performed comparing three levels of complexity in menu design. Subjects consisted of novices and "experts" in computer usage. It was found that novice performance diminished rapidly and statistically significantly once the third complexity level was presented; expert decrement also occurred, although not significantly.

It was concluded that at a certain level of software complexity, operator performance diminishes significantly. Note the characteristics of this conclusion:

1. The validity of the study results was implicitly assumed.

2. It was also assumed that the performance decrement found in the study is characteristic of the population as a whole, not merely of novices or intermediate level operators.

3. Menu complexity is assumed to be equivalent to software complexity in general.

4. The implication inherent in the conclusion was that software designers should and can restrict the complexity of their designs.

All of these extensions of the basic performance data may well be justified. The point is they are extensions, and even though they appear reasonable, the evidence for the extensions is often lacking. On the other hand, bare statistical differences are not intellectually very satisfying until they are related to the population as a whole and software as a whole.

The meaning in the conclusion is supported by what can be called "pragmatic assumptions." Pragmatic assumptions are to be found throughout the measurement process. For example, one almost always assumes that the subjects selected for measurement are valid representatives of the population to which the results will be generalized. It is assumed that the size of the significance of difference statistic (e.g., .01 vs. .05 level) demonstrates the relative importance of the variable involved in the treatment condition. The assumption is made that this variable, when measured in the treatment condition (i.e., in isolation from other variables), produces effects that are the same as when (i.e., in the OE) that variable is part of the mix of all variables affecting a specific performance.

These are pragmatic assumptions because they permit the measurement process to proceed. This might invalidate the measurement if any of the assumptions were refuted. The fact that these assumptions are not challenged or even noted in writing the final report means they have practical usefulness. That is why the correctness of the assumptions is never tested.

There are empirical tests that would confirm the pragmatic assumptions, but they are almost never performed. Further testing, such as rerunning the test with different subjects perhaps, somewhat different stimuli, and so on, is part of the validation process that is generally ignored. Validation is discussed later in more detail. However, the assumption by the researcher that the pragmatic assumptions are themselves valid makes it less necessary to validate empirically.

Writing the Research Report. For conceptual research, the final published report is arguably the most important part of the measurement. As has been discussed elsewhere (Meister, 1999), publication is, outside of research required by the job, the major motivating factor for conceptual research. The importance of publication is much reduced for system-related studies, the final reports of which are also published, but usually do not enter the general HF literature and have a restricted audience.

All reports are sanitized, that is, the problems involved in getting from initial hypotheses to study conclusions are not described. The "discussion of results" section of the published paper allows the author to extrapolate on the basic findings, and to go beyond the conclusions. Moreover, because words are more expressive and have more connotations than data, the researcher's speculations interact with the data. The result may be confabulation, because the speculation is presented along with "hard" data or statistical analysis. As a result, there is a tendency on the part of the reader of the paper to substitute the speculation for the hard facts or at least to commingle them. Fewer professionals read the quantitative data in the paper than read the conclusions and the discussion of results (see chap. 8), and the qualitative material is easier to remember than the data, especially if the data are phrased in terms of abstruse statistical tables.

Authors of papers are encouraged to make the application of their research results fairly specific. Unfortunately, these applications are often simply *pro forma* statements that the results can be applied to one aspect of HF or another, without indicating how the application should be made.

HF GOALS AND THE SELECTION OF RESEARCH THEMES

Chapter 1 began by developing a logical relation between the nature of the HF discipline, its goals, and the purpose of HF research. It was assumed that the research was performed to implement HF aims.

Several goals of HF research were identified: (a) to determine how people respond to technological stimuli, (b) to aid in the development of human–machine systems, and (c) to enable prediction of human performance as a function of equipment–system characteristics.

The fact that theorists have logically identified certain research goals does not mean that HF professionals necessarily accept (agree with) these goals, or are even aware of them. The survey described in chapter 3 will help to answer this question. The next question is: Do these general disciplinary goals exercise an effect on the actual selection of research themes?

There is an inevitable gap between these relatively abstract goals and the individual themes selected. If, for example, researchers decide to study decision making as a function of the amount of information provided, how does this study implement any of the three research goals?

To determine if a relation exists, it is necessary to examine the characteristics of HF research over the years, as the author has done (Meister, 1999; Meister & Enderwick, 2001).

Studying the human response to technology would require a heavy involvement of technology in the research; to aid in system development, much of the research would be initiated by, or deal directly or indirectly, with design questions. For prediction of human performance, some of the research would deal with the development of methods to support quantitative prediction.

A review was conducted of the HF literature over the years 1972–1999 (using as a sample papers published in the *Proceedings of the Human Factors and Ergonomics Society* annual meetings). This review supported none of the expected effects of those overall research purposes. It found a research literature focused on the human (naturally so), with only a tangential concern for technology. For example, Maltz and Shinar (1999) studied the viewing performance of senior citizens, with technological involvement represented by a synthetic driver situation. Technology used in this way provides only a *context* for the human's behavior. Technological context is important, of course, but only indirectly.

Only 10% of the *Proceedings* papers had anything to do with system design. On the other hand, these studies had a striking resemblance to experimental psychological studies. There were almost no papers dealing with quantitative prediction of human performance; in fact, most of the research in this area has been performed by people who are reliability engineers, or, even if they are HF people, are outside the mainstream of HF research activity.

All of this raises certain questions:

1. Why do the overall HF research purposes not exercise more influence on selection of individual research themes?
2. How do researchers actually go about the business of selecting their research themes?

It is relatively easy to hypothesize why the overall HF research goals do not exercise a stronger influence. The abstractness of the research goals makes it difficult for researchers to see a correspondence between the abstract goal and the relatively concrete research theme they select.

There is also the strong influence of the specialty area in which researchers work. Such specialty areas have their own very specific research goals that, in the mind of the researcher, take precedence over higher level goals. Individual researchers specialize in system-related fields, such as aviation and computers, or in function-related fields, like decision making or aging.

Research performed in relation to systems of various types is more likely to be channeled or constrained by the characteristics of those systems. If an individual is, for example, an aviation researcher, then cockpit arrangement is more likely to be the subject of research than something else; computer specialists are more likely to research software displays like icons. There may be somewhat fewer constraints in dealing with human functions like vision or cognition, although even here the nature of the functions reduces the researchers' freedom to study whatever they wish.

Because specific research themes do not derive from overall research goals, the following question arises: How is the actual selection process performed? That process involves a continuing search for information, even outside the actual process of planning a study. All professionals engage in a continuing review of papers describing new research in their specialty area. These papers point out, in their discussion sections, inadequacies in needed information, discrepancies in research results from multiple studies, and obscurities needing illumination. This is the importance of the theory and speculation usually included in these discussion sections. The combination of all the theory and speculation in the mass of published papers may (along with more direct channels such as symposia and paper presentations) serve as the voice of the research community. In very simplistic terms, they may tell the would-be researcher what needs to be studied.

THE ROLE OF VALIDATION IN MEASUREMENT

Validation is not a process restricted to the individual study; it is critical to measurement as a whole. Validation has traditionally been thought of as an attempt to verify the truth of research results.

The first thing to know about validation is that, like the weather, everyone believes validation is essential to HF as a science, but hardly anyone does anything about it.

There are also problems of defining validation and validity. The concept of validity is commonly defined in terms of whether a measurement operation actually measures what it intends to measure. Reliability is associated with validity: If the measurement is repeated, essentially the same results should be derived. Both reliability and validity must be known because a measurement that is reliable is not necessarily valid.

Three types of validity have been postulated: *content, construct,* and *criterion* validity.

These validation indices, which have been applied primarily to personnel measures, are essentially inferential measures; they are based on the definition of a validity "construct" and what is assumed to be the attributes of that construct. It is assumed that if the measurement operations performed adequately represent the construct's varied attributes, using such techniques as internal consistency, factor analysis, and so on, then the measurement is valid.

Of greater interest is criterion validity, which has two types: concurrent and predictive validity. Concurrent validity can be assessed by determining the correlation between the measure whose validity is to be determined and some other measure *known* to be both valid and reliable. Predictive validity determines whether the HF measure predicts some other measure, also known to be valid and reliable. The difficulty with all these measures of validity is that they are essentially analytical and are therefore skewed by the analyst's assumptions and biases. The special difficulty with criterion-related validity is finding a measure (already known to be both valid and reliable) to be correlated with the undetermined measure. If measures are not generally validated, then how is it possible to find an already validated measure?

The HF interest in validation is the extent to which a measurement product actually serves a potential user's needs. In that formulation, the author follows the thinking of Petjersen and Rasmussen (1997). The essential construct in this validation paradigm is *utility*, which can be applied not only to conceptual research (where the problem of validation is most difficult), but also to system-related research. In the latter, however, the problem is less important, because the research is modeled after the system to which it refers.

In any event, the premise is that if research outputs can be applied to satisfy a need of the research "user," then the results are valid per se, because utility is what researchers really want from their measurements, not some conceptual or statistical measure that no one except theoreticians pay attention to.

In the proposed concept, researchers do not validate data or even conclusions; they validate the projected or anticipated uses of the data/conclusions by actually putting them to use. The empirical application of the data/conclusions is the verification (validation) of the original projected use.

Use validation comes closest to predictive validity but does not require correlation. Use validation is closely related to the application of research results to the development of design guidelines. The primary users of the research material are the HF design specialist and sometimes the design engineer.

One application format that the validation takes is the design guideline, which consists of the principles and data used to evaluate the adequacy of the physical interface from a behavioral standpoint. Design guidelines have been utilized since the beginnings of HF.

A less common application format in which research can be validated is to predict quantitatively the performance of system personnel in operating a human–machine system. If the prediction is verified (by measuring the performance of system personnel in actual systems), then the research results that led to the prediction are validated.

The validation process is not an easy one; it requires the gathering of information/data from all the HF research, the selection of those most relevant to system design and use, and the assembly of the data/conclusions to derive design guidelines. The process is exemplified by much earlier work (Munger, Smith, & Payne, 1962).

Such a validation process is not appropriate to the single study (because data from several studies are needed as a quantitative basis for application); thus validation is meaningful only for the totality of all studies relevant to a specific function (e.g., decision making) or a type of system (e.g., process control plants). A preparatory step for the validation process requires the development of a taxonomy of system-relevant HF categories. This taxonomy will serve to organize the research into relevant items.

Nor will this validation process develop a single number representing the *degree* of validation. It seems unlikely that it would be possible to develop a method of quantizing use-validity. In the utilization framework, however, this kind of quantification would be unnecessary. Utility *is*; it does not require a number to represent it. This validation process ties measurement as a HF methodology very closely to system design as a required HF function.

Obviously, criteria of utilization success would have to be developed. In addition, the use-validation process has to be an ongoing, continuing one, because research will not cease.

The design guidelines, serving as the validation mechanism, are relatively molecular. Hopefully, continuing refinement of the use-validation process, linked with research on the molar characteristics of systems, such as complexity and organization, will eventually lead to the ability to predict how effectively system personnel will be able to operate and maintain total systems. This, however, is a much longer range goal.

The use-validity process described is not one that will be achieved immediately. Conservative professionals will not want to change their customary measurement practices (which ignore validity), and inevitably considerable work will be required to assemble, analyze, and transform behavioral data into design guidelines. HF professionals, like everyone else, abhor work.

Any set of scientific abstractions, like relevance, validity, reliability, and generalizability, can assume a number of different meanings over time. Professional attitudes toward measurement phenomena may change and require revision of these meanings. All of this may produce confusion in the professionals' concept structure (see examples in chap. 3). Validity is defined operationally as the comparison of study conditions and results with real people (not selected subjects) performing real tasks with real systems in the operational environment. If these agree, then the study results are valid. Of course, researchers must measure in the operational environment to assure this validity, something most professionals may not be willing to do.

There can, of course, be other concepts of validity: for example, one may be the concordance of new results with previous research conclusions. Validity may also be thought of as absolute truth (in which case it cannot be tested or quantified); or validity may be conceptualized in terms of a comparison of test scores. However, if the HF goal is to assist in the development of new human–machine systems, then performance with these new systems must be related to the research that has been applied to these systems.

THE RELATION BETWEEN OBJECTIVE AND SUBJECTIVE MEASUREMENT

The scientific ethos developed by the "hard" physical sciences like physics and passed on to psychology and then to HF applauds objectivity and denigrates subjectivity.

Although subjectivity is largely irrelevant to physical processes, every human performance that can be measured objectively (including psychophysical processes) is accompanied by subjective performance as represented by the cognitive and imaging activity of the human. This makes the physical science model of measurement somewhat inappropriate for behavioral science.

Because humans think about their performance, that thinking presumably modifies the physical performance. For that reason, even when performance can be recorded objectively, it is a cardinal error not to investigate the cognitive performance that accompanies it. No objective performance should be recorded without concurrently or sequentially recording its cognitive/attitudinal parallel.

A staunch behaviorist would argue that if performance can be recorded objectively, like the winning time in a race, that is all that is necessary. That assumes that one human performance measure is sufficient to describe all behavior and that performance is entirely monolithic. Few professionals would agree to this.

Moreover, there are many performances that are almost wholly cognitive and for which objective performance measures are insufficient. Analysis, decision making, judgments, and so on, cannot be measured objectively unless the subject performs a physical action in parallel that reflects the cognitive process. That physical act (e.g., throwing a switch or moving a mouse) may be completely uninformative. The alternative is to ask the subject for a report.

Relevance to the topic being measured becomes a criterion for the inclusion of subjective data in the measurement record. This applies to physiological data also. Obviously, there are situations in which physiological data are critical, and others in which they are meaningless. In the measurement of astronaut performance, heart rate and body temperature are critical, because they are indices of effective performance, and indeed of safety. In the measurement of sonar detection or command decision performance, on the other hand, the same physiological indices would be irrelevant.

The inadequacies of subjective data are well known. The medium of subjective data transmission (the human) may be biased to present incomplete or inaccurate data. The human is not noteworthy for highly accurate observation. As a consequence, one reason why subjective data are less preferred than objective performance measures is that subjective data are inherently more difficult to collect and to interpret than objective data.

This may seem to fly in the face of normal experience (in real life people are constantly asking questions of others), because it is very "natural" to ask for a self-report. However, it is difficult to phrase a question so that it cannot be misunderstood. It is easy to develop a Likert scale (and every researcher has, at one time or another, done this), but extremely difficult to develop a scale that will not influence and bias the subject's responses. The language in which the subject is asked to respond has so many connotations that the opportunity for misrepresentation and misunderstanding is very great. The subject's interpretation of the questions posed may not be quite what the researcher had in mind. Subjects may not be able to describe completely what they have seen or felt or the internal processes that parallel an objective performance.

On the other hand, however, the data they do present is often unavailable by any other means and may have a unique measurement function: to explain the why and how of human performance when these are not inherent in the objective measurement record. Subjective data may be diagnostic, as when the researcher asks the subject following a test, "Why did you do that?"

In any event, it is impossible to avoid subjectivity in measurement, if only because the researcher is human and may be a biased and less than adequate measuring instrument.

This is not the fault of measurement per se, but of the human agent who performs the measurement. Researchers may think that the experiment disposes of the variance associated with them, but researchers often approach their measurement with biases to the hypotheses they have created for the study.

All these caveats may suggest that subjective performance indices should be avoided, but there are just as many problems with objective data. Unless the stimuli presented are unequivocal, objective data have two faces: the obvious physical action that instrumentation can record; and the reason *why* the subject performed the physical action. Many times the two coincide (e.g., pushing a button when a light becomes green in fact corresponds to the perception of the light); but many other physical actions are only partially representative of the internal processes that initiated that action. The simplest physical action may have the most complex internal process as a parallel. Even pushing the button, as described earlier, may have been accompanied by intense cognitive activity. The objective record of button presses tells nothing about the cognitive activity that accompanied them. Researchers wish to know *why* an action was performed. In a proceduralized task, the rationale for any activity is inherent in the steps of the task. Few tasks are, however, so completely proceduralized. If the researchers have at the end of a study any unanswered questions, then the objective study data have been insufficient. All studies may be insufficient; otherwise, why is there extensive speculation in every paper to "explain" what really happened, after the objective record has been examined?

Researchers should never be satisfied with the first glance at their data. The key to successful measurements is always to ask questions (e.g., What does the performance record mean? What were subjects thinking about when they did what they did? Were there any opportunities for misunderstanding on the part of the subjects? If the data reveal an unusual performance effect, what does it mean? How many different explanations of the phenomenon are possible?).

APPLICATION AND UTILITY IN RESEARCH

All the research in the world is meaningless unless it can be applied. The only exception to this argument is if research is considered an art form; art has no rationale except to give pleasure.

Researchers would say, if pressed, that the purpose of research is to gain knowledge and, in Bacon's words, knowledge is power. Because knowledge can never be complete, it is necessary to continue researching.

If this argument is adopted, then the question becomes: Knowledge of what? Is knowledge monolithic, so that whatever is being studied is worth studying? This seems unlikely. Knowledge can be divided into the useful and the useless, the important and the trivial. If my newspaper prints the temperature in Beijing, China, and I read this, I have gained knowledge, but the information is useless to me, unless I am contemplating a trip to Beijing.

But, it will be objected, I am only one person. Even if I am not interested in Beijing, others may be and may wish to know how warm it is in that city. This is a reasonable objection, but it leaves a problem. The communication of all knowledge of all topics is an impossible task, even if the Internet is available for help. The human lifetime is limited and our resources are finite, so that, like it or not, some choice of knowledge and information must be made.

Researchers must decide on what is more or less important, because logically they would choose the more important. The criterion of importance is not made solely by the individual researcher, it is made by publications (what is published is presumably important) and by funding agencies (what is financially supported is important). These criteria are both objective (i.e., they can be observed in action), but they are also subjective, because it is a subjective judgment of experts as to what is funded and published.

Can the judgment of what is important be mistaken? Very possibly, because these are judgments of individuals. This can be avoided, however, by using the criteria of *applicability* and *utility*.

There can be no error in applying the judgment of applicability and utility, because these can be observed in terms of the concrete consequences of the knowledge (e.g., a change in a human–machine system or a new method of training).

A further distinction must be made between information and knowledge. Information may be no more than an assembly of facts, as in an almanac. These facts may or may not be useful; it depends on the questions asked of the information. The most descriptive illustration is a game called "Trivial Pursuit" in which players are required to answer questions with individual facts having no relevance to anything except the question. It could be said that the information is useful (for that question), but has in effect no larger context.

If knowledge has any attribute at all, it is that it has *value*. The essence of that value is that something can be done with that knowledge; something that is important can be explained, such as how to cure cancer or how to design a human–computer interface. On this basis, it is possible to discriminate between HF research that is more or less important, more or less useful.

Researchers dislike this distinction because when it is asked about a particular research question or study, it creates uncertainty; others (e.g., funding agencies) may decide, regardless of the researcher's interest in a theme, that it is not worth studying.

It would be much simpler if the question of research value were never raised. Then every researcher could study anything at all, without any self-doubt. The only problem is that there is not enough money to support this research, or, if it is actually performed, to publish it in HF journals.

Value judgments are ubiquitous in research, but professionals do not wish to admit this, because to admit this is to admit that some research has more value than other research, which means that research can be measured and evaluated on some scale of utility, and that research can be *directed* (advised, forced) to some themes rather than to others by applying value criteria. This diminishes the researchers' freedom to study what they wish, but because research requires funding, the funding agency ultimately decides what will be studied. Researchers hate to admit this, because they now have to deal with questions, such as: What is the basis of making judgments of value? Who makes these judgments, and are they entitled to do so? How is utility defined? How can HF research be made more valuable?

There are two types of utility: *conceptual utility*, whose function is to stimulate further research, and *physical utility*, which assists in the design and use of human–machine systems. Conceptual utility is much more difficult to identify than physical utility. The effects of conceptual utility are not immediately ascertainable. Physical utility, when it occurs, is less obscure, although obscure enough. Conceptual utility has no specific behavioral reference; it can be applied to any research content. Physical utility can be applied only to dynamic phenomena (e.g., the human–machine system and human responses to technology).

HF value should relate to the HF purposes specified in chapter 1. Whatever research implements, these purposes (understanding human responses to technology, aiding system development, and predicting human performance) should have more value than any research not specifically implementing these goals.

Finding utility in research is much like old fashioned gold mining procedures: sifting through thousands of papers, as the originators of the DataStore (Munger et al., 1962) did, to find a few that provide useful data. Each specialty area will have its own set of themes to use as a sifting device. In the case of system development, the following themes can be used to discover potentially useful research: How can behavioral data and principles be transformed into design guidelines? How do system attributes, notably complexity, affect operator performance? What kind of information is needed by HF design specialists? What are the parameters of the design process?

If there is agreement that HF research should be applied, then the questions become: Applied to what? And, who should be responsible for making the application? The physical entities with which HF is involved are human–machine systems, and so research should be directed to them. If the application is to system development, then it is the responsibility of HF design specialists to make this application, because that is or should be one of their major roles.

Even so, researchers themselves also have part of that application responsibility. The one closest to the original research has an advantage in suggesting what should be done with that research. Nor should the application be in *pro forma*, overly general terms, such as, "This research can be applied to system development, because it contains information about how operators respond to information produced by machine stimuli."

THE ROLE OF THE OPERATIONAL ENVIRONMENT IN MEASUREMENT

The operational environment (OE) is an important construct, because in essence it is the operational definition of HF reality. When people talk about "the real world" they refer to the OE.

The OE is the setting, the context, in which the system and the human who is a subsystem of that system perform as they were designed and (in the case of the human) instructed to perform. That setting varies with the type of system (i.e., an aircraft has a different OE than that of a process control plant).

Although the OE is a context for system performance, it is also a system in itself, which means that it is an organization of personnel and equipment pursuing activities directed by a goal. It may seem strange that a setting or context is also a system, but that is because the OE contains its own infrastructure and subsystems. For example, the overall OE for a commercial aircraft is the sky, but it also has subsystems consisting of the landing field, the air traffic control station (ATC), the terminal (for its passengers), and so on. The OE therefore consists of a hierarchy, ranging from the molar (in the case of the aircraft, the sky) to the molecular (the baggage carts on the tarmac).

Obviously, the nature of the system under consideration defines the OE for that system, as in the aircraft example. If the intent is to study the ATC (e.g., its functions) within the aircraft OE, then OE must be considered while measuring the performance of the ATC. At the same time, the ATC has its own *physical* OE, consisting of radar consoles and communications equipment, office furniture, and so on, and a *behavioral* OE consisting of stimuli affecting the controller procedures they must follow, and so on. Each subsystem within the OE has its own OE, its own setting. There are, then, multiple OEs, each of which serves as context for some human performance and for measurement of that performance.

Researchers can therefore decompose or dissect the OE, just as they can any other human–machine system, by deciding what part of it is to be measured.

The OE has certain interesting characteristics; for systems that move through space, such as a ship or an aircraft, the OE changes its characteristics rapidly; for example, the OE for an automobile on a cross-country trip has many geographical sub-OEs. Manifestly, the OE as a construct is tremendously complex, although its physical manifestations are much simpler, because the concept of the OE can be restricted to a very small part of the larger OE. A ship can be considered as a whole, with its overall OE, or concentration may be on the ship's bridge.

The importance of the OE for HF measurement is that to make any of the measurement outputs meaningful, the measurement must be made in the OE or must be made in a venue that simulates the OE; and, in any case, measurement outputs must be referable to the OE. That means that any measurement output must say something about the system and/or the humans as they ordinarily function within the OE. Interest in human–system performance is on what goes on in reality (normally, routinely), not in a laboratory or even in a simulator (unless the simulator closely replicates the OE).

Chapter 1 stated that the OE is the ultimate referent for the HF discipline. Professionals cannot object to this, because it is not only a conceptual assumption (a construct), but a physical one; what individuals see when they look around, what they touch or hear, is the physical manifestation of the OE, its concrete expression. Their home is an OE; when they go out on the street, the street is another OE. The number of OEs is not infinite, but there are many of them, depending on the range of activities in which the system (including the human) engages.

What does this have to do with HF measurement? It means that when initiating research or looking at its outputs, researchers must always ask: What will or does it tell them about how the human–system configuration will function in its appropriate OE? The researchers and users of the research are not interested in how the system or the human functions in an ideal situation like the laboratory or a simulator or a test site, but how will it function within the OE?

In consequence, the OE provides criteria, the most important one being the criterion of relevance. If a measurement/research cannot be related one way or another to the OE (which means some human task and system), it has very little significance for HF. This criterion can be utilized even before a study is initiated (in determining test conditions), as well as afterward, in relation to its outputs.

The Conceptual Structure of the HF Professional

Chapters 1 and 2 described a measurement process (Tables 1.1 and 2.1), which is the way measurement (research) is commonly performed. These are customary measurement practices, henceforth referred to as measurement practices.

CHARACTERISTICS OF THE CONCEPTUAL STRUCTURE

It is reasonable to assume that these measurement practices are directed by the researcher's higher order assumptions, concepts, beliefs, and attitudes; these can be termed the *conceptual structure* (CS) of the HF professional. This chapter explores that CS to see how it interacts with and influences measurement practices.

Most of the CS discussion is theory. Measurement practices are overt and are described (although inadequately) in published papers. There is no empirical evidence about the professional's thinking about the problems encountered in these practices. To get some of the needed evidence, a survey was performed (which follows this section).

It might appear as if the CS and measurement practices were separate and distinct, and in large part they are. It is possible, however, to conceive of the thinking operations of the CS as having two levels, like a house: the first, a higher order, fairly abstract level consisting of concepts, assumptions, and beliefs; and a lower level, like a basement, of measurement practices, much more concrete and mundane, which deals with specific measurement problems as they actually occur in the real world.

Although the practices described previously are relatively concrete (see Table 2.1), they do involve some complex conceptualizations (e.g., the development of testable hypotheses). It is legitimate for this reason to consider these practices as part of the CS. The upper story of the CS can be thought of as a motivator and evaluator of the professional's measurement practices; the upper level includes scientific ideals, measurement goals, and in particular the criteria that can be used to evaluate the adequacy of the measurement effort. The basement level practices represent how the professional solves actual problems in interaction with the higher order CS.

The two levels are always in tension, always in some form of conflict, because the ideals of the upper CS cannot be easily or fully accomplished. Nontechnical factors, such as funding and researcher weaknesses, make it difficult to implement the scientific ideals. It is, moreover, not easy to deal with the higher level CS because its concepts are quite abstract; this makes it difficult for the professional to translate them into concrete operations.

Unable to apply scientific ideals fully in their measurements, because reality constraints repeatedly interfere, HF professionals make use of pragmatic assumptions to reduce the tension between the two levels and to rationalize practices.

One higher order ideal is to perform HF research in accordance with the general aims of the discipline—to promote understanding of the human–technology relation and to assist technology through provision of principles and data to system design. Another scientific ideal deals with what the characteristics of HF principles and data should be (e.g., relevant, valid, reliable, generalizable, useful, etc.).

A case can be made that professionals must continuously ignore these ideals, because, except in extraordinary circumstances, ideals cannot be fully implemented and often must be ignored; this is what produces the tension between the two CS levels. Reviews by Meister (1999) suggested that HF research does not fully analyze the relation between humans and their technology, but instead concentrates on the human reaction to technological stimuli. The two are not quite the same, although they may appear so. The relation between the human and technology is much more than the former's reaction to technological stimuli. Of course, humans *react* to technological stimuli, but humans also *create* the technology they respond to through design and engineering. To thoroughly fulfill HF goals, it is necessary to study design processes.

Another factor producing tension is that what HF studies is determined not so much by the HF research community as by the willingness of agencies to fund particular questions. Funding agencies do take the HF research community into account in making their decisions, but often these decisions are made for nontechnical reasons.

Higher order CS ideals mandate validation of research conclusions. However, the validity of most HF research results is unknown and therefore suspect, because empirical validation efforts are not performed.

Moreover, the supply of appropriate inputs from research to influence system design is halting and weak, because that supply is only a *byproduct* of experimentation.

The generalization of research results to operational systems and their personnel is unknown, because generalization is automatically assumed by researchers.

From this, it can be deduced that higher order CS requirements are routinely ignored in the performance of measurement practices. This results in part from the unwillingness or incapability of professionals to change the test situation. The professionals' recognition that CS ideals are ignored adds to the tension.

Unfortunately, not enough is known about the professionals' CS. Certain parameters listed in Table 3.1 and discussed in the next section are probably involved, but how professionals feel about these parameters is unknown.

CS Parameters

The *basis of HF research* (Item 1, Table 3.1) in large part determines the nature of the research themes professionals select to study. These themes are directed by HF *research purposes* (Item 2), as well as the special characteristics of the specialty area within which the professional is working (e.g., aero-

TABLE 3.1
Parameters of the HF CS

1. The basis of HF research
2. HF research purposes and specialty areas
3. The distinctiveness of HF research
4. HF research constraints
5. Basic and applied research
6. Data transformation
7. Subjective data
8. Validation
9. Prediction
10. Research utility
11. Research relevance
12. The importance of control
13. Measurement venues
14. Research effectiveness
15. The research community
16. The store of knowledge
17. Speculation and theory
18. Customary measurement practices

space, computers). Related to Item 1 are the distinctive *characteristics* of HF research (Item 3), particularly its differences from psychological research (i.e., its emphasis on technology and systems). What can be done in accomplishing research purposes is affected by *constraints* (Item 4) such as time and funding.

Parameters 1–4 lead to a *distinction* between basic and applied research (Item 5), which is also instrumental in selecting research topics and the publication of resultant papers.

The outputs of a study are data, but more important than data, which professionals use infrequently, are the conclusions derived from those data. This requires another parameter: *data transformation* (Item 6), from raw data to conclusions. These data are of two types: objective and *subjective* (Item 7). Objective data being preferred in accordance with scientific principles, the question of how to deal with subjective data is an important one.

The previous parameters describe measurement ideals and processes. Parameters 8–11 involve measurement criteria. For example, if professionals are concerned about the "truth" of these data, then they are concerned about *validation* (Item 8). Although very little is actually done in HF to validate research conclusions, the topic remains a source of irritation to professionals; researchers should validate, but validation is expensive and a nuisance.

The rationale for measurement is *prediction* (Item 9), because measurement outputs are supposed to be applied in a situation other than that in which testing occurred. To use a very simple analogy, if a piece of lumber is measured to construct a table leg, this "predicts" that the lumber selected will be sufficient for the table leg. Prediction (e.g., where one drives to reach a destination) is a common aspect of human functioning. Prediction becomes much more complex, however, than table legs and driving when systems are involved.

It is assumed that research has *utility* (Item 10); otherwise, professionals would lack interest in that research. How is utility defined, however, and whose utility? These are questions of some interest to professionals.

Linked to utility is the criterion of research *relevance* (Item 11) to some referent to which researchers wish to apply research outputs. Application requires relevance.

Adequate measurement requires *control* (Item 12), because control enables researchers to structure the measurement situation in order to answer their questions. Measurement occurs in a particular situation known as a *venue* (Item 13) (e.g., laboratory, test site), which influences the outputs secured.

Measurement outputs can be evaluated on the dimension of *effectiveness* (Item 14) (i.e., that some research is more adequate, more important, more relevant and useful—select your own definition—than other research).

HF research is performed by professionals who form a *community* (Item 15) whose interests overall have a great deal to do with the individual researcher's determination of what to study. The researcher is unlikely to select a study topic that is too far removed from that which the community considers acceptable.

Basic, fundamental, substantive research (readers can choose the adjectives they prefer) is supposed to contribute to something called the *store of knowledge* (Item 16). Because the latter is one of the implied science principles that serve as a rationale for most conceptual research, it is assumed to be important to the professional.

Speculation and theory (Item 17) are critical functions of measurement, but how important are they to professionals?

The last major item (18) is *customary measurement practices.* In contrast to the previous abstractions, most of which are scientific ideals, customary measurement practices are relatively concrete, habitual processes used by the individual in specific situations to solve specific measurement problems. A preference for the experiment, rating scales or interview, or the number of subjects ordinarily utilized in work, are examples of such practices, which are largely independent of the other, more intellectual concepts.

All these parameters are interrelated, but to varying degrees; the professional may be more aware of some of these than of others. For example, customary measurement practices exercise a greater effect on actual testing than concepts of research purpose, validation, and so on, which are elicited in the CS only when they are subjects of intellectual debate.

The purpose of the following survey was to discover how the CS parameters exercise their effect.

THE SURVEY

A mail survey was adopted to investigate the CS. A survey may appear to be a weak mechanism with which to tap so complex and obscure a matter as measurement concepts; yet, no other alternative presented itself. Personal interviews with respondents, with the opportunity to question and investigate responses further, would undoubtedly have been preferable, but the money and time needed for such an effort were not available.

The following is an example of a survey item:

The concepts of basic and applied research have no effect on the way in which HF research is performed.

Agree | | Neutral | | Disagree

The previous scale has several distinctive features. Although it is a 5-point Likert-type scale, it contains both ends of the agreement–disagreement continuum. The scale ends (agree, disagree) imply complete agreement and complete disagreement. Respondents were told to check their responses anywhere along either the agreement or disagreement continuum. This indicated the extent to which they agreed or disagreed with the statement. The distance between Neutral (0) and each end of the scale can be subdivided into 100 points, although response frequencies were tabulated only at 5-point intervals (i.e., 100, 95, 90, 85, etc.).

The survey instrument was mailed to 72 HF professionals in the United States, with 63 responses (response rate of 87%). The sample was highly selected; the survey was sent to senior members of the discipline, most of whom had achieved high status in the profession. The sample included presidents of the Human Factors and Ergonomics Society and the International Ergonomics Association, heads of governmental and consulting agencies, authors of well-known textbooks, editors of HF journals, and university faculty members. The sample included a small number of younger people. The mean for years of professional experience was 29, with a range from 5 to 50 years. Nine respondents had over 40 years' experience. Senior people were targeted because these were most likely to be responsible through their books and presentations for CS ideals and measurement concepts.

Data Analysis

The data were frequency of responses to points along the scale. For analysis purposes, the scale was divided into subsections as follows: Major points on the scale were 100 (complete agreement or disagreement with the proposition), 75 (strong agreement or disagreement), 50 (moderate agreement or disagreement), 25 (weak agreement or disagreement), and 0 (neutral, or no agreement or disagreement).

Respondents were encouraged in their instructions to indicate if they had difficulty understanding an item. They were also asked to comment on the items in the margin of the survey form. These comments were factored into the author's interpretation of the results.

Frequencies for responses to the agreement scale were compared with those for the disagreement scale. The number of responses to zero (neutral) was also tabulated. These values were interpreted as follows. A total (summed) frequency for either agreement or disagreement of more than 50% of the sample indicated general approval or disapproval of a survey concept. A frequency split roughly down the middle of either agreement or disagreement indicated two alternative concept interpretations. A small frequency that disagreed with the general response was designated a minority position.

A more detailed statistical analysis of the data was possible, but the author felt that the complexity of the concepts examined warranted only general statements. Elaborate statistics may tend to create a false impression of definitive fact, when what is actually being reported is only possibilities and conjectures.

Because a few respondents refused to respond to individual items, *N* for these items was less than 63, but never less than 58 or 59.

The 52 concept statements in the survey represent only a subset of all the possible concepts that can be derived from measurement elements, but were considered to be the most important.

Results

Responses to the individual items were grouped into subsections under general themes describing the common nature of these items. The individual survey items are listed at the beginning of the subsection, together with the frequency of agreement (A), disagreement (D), and neutral (N) responses for that item.

The Basis of Human Factors Research.

16. The nature of the relationship between the human and the machine is as much a *fundamental* research issue as are the parameters of any other basic research topic. (A-59; D-2; N-1)

Respondents almost unanimously agreed to this item. Nothing else should have been expected, perhaps, considering that the statement is fundamental to the HF discipline.

When examining the concept in detail, however, it is apparent that it can have multiple interpretations; and this paradoxically favors its acceptance by the respondents. (However they study the human–machine relation is acceptable because that relation is phrased in such general terms.) The professionals' concept of the relation can then be easily adapted to the limitations of the actual measurement situation.

The human–machine relation may be interpreted in many ways: as the study of the human operation of individual machines, like a lathe, or of total systems, like processing plants; it can be related to the use of small hand tools or appliances, or to environmental effects on motorists, or to the decision making of participants in business classes, and so forth. Even more generally, the relation can be defined as the study of the human reaction to technology in general (although few approved when this concept was presented in another item).

In this relation, there are two elements: the human and the machine–system, and where more emphasis is placed on one rather than the other, the whole tenor of HF research can be changed. The point is that until the concept is defined more concretely, with qualifications and limitations, it can mean everything or nothing.

The lack of an operational definition—What does the human–machine–system relation really mean?—suggests that the interpretation of this relation is determined by the immediate needs of the research situation. If a research job requires studying the anthropometry of motorcycle helmets or the effect of street noise on pedestrians, either of these themes can be accommodated by the definition, or, indeed, anything else one wishes to study.

The issue of convenience is also important. With almost any interpretation of the concept acceptable, that interpretation will be selected that is easiest for the professional to deal with. If it is easier to focus on the human rather than on the human–system relation, the former will be the preferred mode of research. Moreover, as the nature of the real-world measurement situation changes, it is likely that the researcher's interpretation of applicable concepts will change to fit the specifications of the measurement problem.

For various reasons (primarily the lack of an operational definition, but also because of the allure of convenience, the researchers' prior training, and their work experience), the concept does not actually direct the research. It is the measurement problem that researchers select (or are required to study), which influences how the concept will be applied to the problem.

The Results section began with this concept for two reasons. It is fundamental to HF, and it illustrates very vividly what the reader will see in other concept responses: That is, how the researcher defines the concept determines the use that can be made of that concept.

The Distinctiveness of HF Research.

23. There are no significant *differences* between *HF* and *psychological* measurement questions. (A-6; D-53; N-2)
36. HF measurement *differs* from all other behavioral measurement because it focuses—or should focus—on the human/machine *system* relationship. (A-49; D-10; N-3)

The two statements focus on the distinctiveness of HF research. Item 23 suggests that there are *no* significant differences between HF and psychological research. This concept was substantially refuted, although there was a small minority opinion. Item 36 suggested that the distinctiveness of HF research depended on its system relation. This was agreed to, but again with a substantial minority.

Although researchers believe there are important differences between HF and psychological research, it might be difficult for them to pinpoint the differences. In Item 36, the author offered the system, and the human relation with the system, as the distinguishing difference, but a substantial minority rejected that viewpoint. There may be those who reject the system as being critical for HF.

The minority opinions in these two items raise significant questions. If the system is not what is distinctive about HF, then what is? If there are significant differences between HF and psychological research, then what are they?

The belief in the differences between HF and psychological research does not, unfortunately, lead to distinctiveness between HF and psychology in actual research.

Human Factors Research Purposes.

27. One major *function* of HF research is to *assist* in the *development* of human/machine systems by applying that research to design. (A61; D-0; N-1)
41. The major *purpose* of HF research is to discover how people *respond* to and are affected by *technological* stimuli like display characteristics. (A-27; D-29; N-5)
50. Most researchers do not think of the overall *purpose* of HF research when they *select* a topic for study. (A-38; D-10; N-12)
52. The main reason for performing research is to remedy a lack of needed information. (A-50; D-10; N-1)

Although the common theme in all these items is the purpose of HF research, each item suggests a separate purpose: Item 27, to aid system development; Item 41, to discover how people respond to technology. Item 50 suggests that in selecting a research topic professionals do not think of any overall HF purpose. Item 52 suggests that a major reason for performing HF research is the need to fill a gap in information.

All respondents agreed with Item 27. They had to agree, because many HF specialists do work in system design. Moreover, there was a very high level of support, with many respondents agreeing at the 100% level of belief.

Understanding how people respond to technology (Item 41) is a much more general goal than aiding system development; this may be the reason for the almost complete division of opinion. However, if technology is thought of (e.g., in the form of human–machine systems) as providing stimuli, human performance can be viewed as a response to those stimuli. The half of the respondents who rejected the notion may have objected to this being the major purpose.

Respondents agreed generally, but not overwhelmingly, that researchers do not think very much about any overall HF purpose in selecting their research topic (Item 50). This break in the chain linking the overall conceptual structure (HF goals) with customary measurement practice is disturbing, because it suggests that the higher level CS is relatively ineffective in influencing professional behavior.

If the choice of a research topic is not determined by overall HF goals, then it must be because the specialty area (e.g., aerospace, medical devices) is the researcher's primary concern. The job requirement may be important here also. If, for example, an individual works for an agency that supports research on aging, then the agency wishes to know more about aging, and any researcher's personal interest, in alcoholism, for example, must be ignored unless the researcher can convince the agency to study alcoholism in the elderly. Generally, the research community (whether as individuals or as a group) simply assumes that whatever research is performed will in some fashion aid the overall HF purpose.

For Item 52, there was general agreement that the purpose of performing research is to fill in gaps in information. It seems that this concept would be acceptable to all, but a minority disagreed, despite the fact that the need to acquire information or to fill informational gaps is a common justification in published papers.

But, why should there be completely divergent responses to the same concept? The author suspects that it is impossible to interpret abstract concepts without researchers referring these to their previous experience with measurement problems. On that basis, there may be differences in agreement or disagreement and variations in the strength of a belief in a concept, because respondents have had different experiences with real-world measurement problems.

Human Factors Research Constraints.

4. Science is most effective when the scientist is completely *free* to study what s/he wants to study. (A-24; D-26; N-12)
9. For most professionals the *nature* of the job almost completely determines what and how they can research. (A-46; D-10; N-5)
21. Reality *constraints*, like money, subject availability, job requirements, are important, because they influence *how* the researcher performs his/her research. (A-61; D-0; N-1)
30. *Limitations* on how one performs HF research/measurement depend in large part on *funding*. (A-54; D-6; N-2)

All the aforementioned items are related to the central problem of nontechnical factors that impact measurement.

Somewhat more response uniformity for this common theme was expected, because these nontechnical factors are commonly used as an excuse for research inadequacies.

The abstractness of the concepts in Items 4 and 9 may have increased response variability. It is noteworthy that when related concepts are relatively concrete (Items 21 and 30), response unanimity is sharply increased. HF professionals, like most people, do not do well with abstractions. When concepts are abstract, their meaning is determined by the more concrete experiences of the individual. Thus, abstractions can be compensated in part by relatively objective operational definitions of the concepts. If the concept "technology" is defined as referring to human–machine systems, some of the vagueness of the concept disappears. Then, although the concept will still be partially determined by past experience, the range of the concept connotations will be narrowed by the definition.

The theme of Item 4 is freedom and its effect on research efficiency. The split in responses to this item may perhaps be related to the influence of the funding agencies. Those who have done well with the funding agencies may not require freedom to perform research; those who have done less well may believe that they would do better without the agencies. The unanimity of response to Items 21 and 30 indicates how important money is to research.

Basic and Applied Research.

28. There are significant *differences* between *applied* research questions and those of *basic* research. (A-43; D-14; N-5)
44. Most professionals would *prefer* to perform *basic* rather than *applied* research. (A-7; D-38; N-15)
45. Basic research is defined by the need to discover how invisible mechanisms (*variables*) produce/*affect* visible performance. (A-28; D-23; N-9)

The distinction between basic and applied research is often found in measurement texts (e.g., Weimer, 1995). Although there may be difficulty in determining the precise differences between them, they do represent a common conceptual dichotomy. It is possible that concepts that cannot be operationally defined are in the nature of "myths."

Note, however, the minority negative response to Item 28 (which suggests there are no significant differences between basic and applied); and the clear split in response to Item 45, which suggests a certain ambiguity in the definition of the distinction. Most respondents reject the notion that basic research is preferred over applied (Item 44); there is, however, great ambiguity about the idea, as shown by the large neutral response to this item.

The author has an operational point of view: that system-related research is largely applied, whereas conceptual research (even when it treats molecular system components like menus) is what is thought of as basic. But, this orientation is probably too simplistic.

A frequently expressed viewpoint is that research that contributes to knowledge is basic and research that does not is applied. However, many professionals believe that all research (however poor its relevance and sophistication) aids knowledge. This knowledge-based distinction is not useful, because, as becomes evident later, no one really knows what the concept of knowledge means.

Another way of viewing the basic–applied distinction might be in terms of what can be termed "data worthiness." If a research produces data and principles that deserve to be included in the discipline's data archive, then it should be considered basic; if not, the research is not basic, although it could hardly be called applied. But what is the criterion for inclusion in the archive? This remains to be determined. However, application implies usage, and research that does not produce important material can hardly be applied to real-world problems.

All this can be summarized by asserting that the basic–applied distinction is irrelevant to research, except possibly to theoreticians who use it (and the basic = knowledge relation) to fend off questions about the utility of their work. To simple-minded folk it may appear that what is needed are data and principles to assist the development of a "human-friendly" technology.

The bottom line is that if the distinction cannot be operationally defined (and it appears it cannot), then the distinction is unwarranted. The distinction has no positive consequences, but it does have a negative one: It clouds the researcher's mind and makes it more difficult to select appropriate research.

Data Transformations.

8. Study conclusions should simply state in words what the data reveal in numbers; should never be more sweeping than the numbers permit. (A-31; D-30; N-1)
13. Measurement conclusions should say only what the data tell one. (A-27; D-30; N-5)
20. The fact that one has to translate data into words to describe conclusions *distorts* to some extent the meaning of the data. (A-30; D-26; N-6)
35. *Written* discussions of what is known about a particular topic often *intermingle* theory and speculation with facts and conclusions. (A-52; D-2; N-7)

42. *Data* do not clarify issues or solve problems until the researcher thinks of those data as describing someone performing a *task*. (A-36; D-16; N-8)
46. The development of conclusions from data requires the researcher to *transform* numbers into a visual *image* of the *task* represented by those numbers. (A-36; D-16; N-8)

Arguably, the transformation of research data into conclusions and into further speculation and theory is what gives meaning to a study. The process requires, however, a change from results (numbers, usually) to conclusions (words). The words, or rather the meaning implied by the words, not the numbers, are what are remembered and exercise research influence.

The central questions are whether the study conclusions distort or permit confabulation of the data (what *actually* happened) with the conclusions (what the researcher *thinks* happened), and with speculation (the implications of data and conclusions).

In Items 8, 13, and 20, responses were almost completely divided between agreement and disagreement. Almost all respondents agreed with Item 35, which suggested that data and speculation are often conflated. This means that readers may imagine that the speculation is at least partially factual.

Reactions to Item 42 (the task as the meaning context for the data) were mixed, although there was greater agreement than disagreement, with a number of indecisive neutral responses (8). Researchers may not think of their results in usage terms, because data are often viewed by researchers simply in terms of statistical significance and not in terms of the actual performance of live people.

Item 46 (conclusions are assisted by a visual image of the task) also produced dichotomous responses, but with an even larger number of neutral viewpoints (12), suggesting that these respondents may not have understood the parameters of the concept.

Subjective Data.

1. Performance data cannot be fully understood without *interviewing* test subjects to learn their experiences (understanding) of what happened to them during the study. (A-42; D-18; N-2)
14. Data secured by subjective methods, such as interviews or questionnaires, are *untrustworthy*. (A-6; D-52; N-3)
49. *Subjective* methods, such as questionnaires, are, despite their inadequacies, *necessary* to secure information directly from people. (A-58; D-3; N-1)

Responses to the three items dealing with subjectivity are largely in agreement that, however anyone thinks about subjective data, they are necessary, because they add to understanding of objective data. Respondents substantially rejected the notion that subjective data are untrustworthy. Nevertheless, the existence of a minority viewpoint, however small, testifies to the lingering suspicion that many researchers have about subjective data.

All research, and all experimentation, even that which centers on physiological responses, involve subjective elements, because humans are involved in all research phases.

What researchers may object to in subjective data is the lack of researcher control over the data collection process; the subject largely, but not completely, controls that. In "objective" data collection, subject performances are subjective, of course, but the control imposed by experimental treatment conditions is assumed to mitigate the subjective elements.

Self-report data may be flawed by self-interest, but they are nevertheless data that can be useful, if interpreted properly.

Some professionals may also fear that, if objective data is gathered about a performance and such data are compared with subjects' perception of their own performance, there may be a conflict between the two sets of data, and then what is to be believed?

Validation.

6. *Validation* of any set of study results is *necessary* for *acceptance* of those results as being "true." (A-54; D-7; N-1)
10. Validation of research results is not particularly *important.* (A-2; D-58; N-0)
32. *Lack* of *interest* among professionals is the main reason why *validation* studies are rarely performed. (A-15; D-34; N-11)

It has already been suggested that validation is important because every performance contains a certain amount of error with relation to the question asked about that performance. Obviously, any single study is only a sample of all possible performances of concern, and therefore can only be partially representative of the whole. However, most researchers in every discipline do not allow this to disturb them unduly.

Responses to the previous items produced very great agreement that validation is necessary and important, even if researchers do not know what they are talking about when they use the term. Respondents strongly do not believe that it is their own lack of interest that is responsible for the lack of validation efforts. The culprit most often accused in the failure to validate is a lack of funding. However, the absence of concern in professionals for lack of validation efforts may also stem from acceptance of the fact that truth can only be approximated. If the full truth is ultimately unobtainable, then why worry about validation?

Another possibility is that validation in the minds of funding agencies describes old material, and the preoccupation in American culture (which includes its research) is for the new, the original.

Prediction.

25. It is impossible to *predict* human performance because behavior is so variable. (A-2; D-58; N-1)
48. Another major *function* of HF research is to *predict* quantitatively the effectiveness of human performance as a function of equipment/system characteristics. (A-61; D-0; N-0)

One of the indices of a mature science is that it is able to predict performance and phenomena of concern to it. One function of theory is to predict performance. Consequently, one of the major functions of HF is to predict the performance of personnel working in a system context.

The two previous items attempted to determine whether HF researchers agreed with this position. Item 25 attacked the question negatively, that because human behavior is so variable, human performance prediction is impossible. This point of view was rejected almost unanimously by respondents.

The concept that prediction is a major function of HF research was unanimously supported. Why such unanimity when other concepts had minority responses? Prediction could be supported, because it is a goal, an ideal, not a requirement, not a specific measurement practice, and, therefore, not truly binding on the professional. Respondents may also have felt that every research conclusion contains a prediction: Under the same or similar circumstances as that involving the study from which the conclusion was derived, the same performance results would occur. This kind of prediction, which is quite general (because most research conclusions are quite general), is not very useful.

HF prediction should be quantitative and related to a specific type of human performance linked to specific system and task characteristics. Such a prediction is comparable to a prediction of the life cycle of an equipment component (e.g., a light bulb of a specific type will have 1,400 hours of use).

Quantitative prediction requires a whole program of research before it is possible to develop the predictive capability (see Meister, 1999). It requires a concept of what researchers wish to predict and the development of a taxonomy of human and equipment operations, on the basis of which archival data can be extracted from various sources. The compilation of behavioral data also depends on having a model of the interaction between the equipment and human performance. Obviously, human performance prediction requires a very extended, continuous effort, and cannot be phrased in generalities.

The extent of respondent agreement on prediction suggests that a strong conceptual belief need not lead to any constructive efforts to implement that belief. Although there have been efforts over the past 50 years to develop human prediction methods (Meister, 1999), these have not involved most HF professionals and there has been little HF interest in prediction. Consequently, what is seen in the prediction concept (as in validation also) is what can be termed a "goal-concept." This means that such a concept has no practical measurement consequences, but remains merely an ideal—a not-as-yet-realized goal. The failure to implement these goals contributes to the tenuousness of the concept structure.

Research Utility.

5. Research which simply *explains* why something happened has only *limited* usefulness. (A-20; D-38; N-4)
15. HF research does not really provide behavioral *guidelines* for design of human/machine systems. (A-15; D-42; N-5)
24. It is not the researcher's *responsibility* to determine what the application of his/her research should be. (A-8; D-49; N-2)
29. Whatever *research* is performed should be *applicable* to and *useful* for something beyond the performance of further research. (A-43; D-14; N-5)
47. Most design *guidelines* (e.g., use language consistently) result from common sense, not from HF research. (A-18; D-38; N-5)

These five items ask in various ways: What does the research community want to do with its research? Research utility is almost never considered in measurement texts. The discipline simply assumes that research must be performed. Whether that research has value, what kind of value, and how to enhance that value are topics about which HF writings are largely silent.

Item 5 suggests that there must be something beyond mere explanation involved in research utility. Explanation is useful and may contribute to the store of knowledge. However, explanation need not necessarily lead to any concrete actions. This concept in Item 5 was rejected, although a substantial minority accepted it.

If research is to aid system development, then it should do so by providing design guidelines. Item 15 suggested that HF research does not really supply these guidelines. This proposition was rejected by a ratio of almost 3 to 1. Respondents believe that research does provide design guidelines, but these must be supplemented by logic, common sense, and experience. Unfortunately, nobody knows how much research contributes to design guidelines, because this question has rarely been explored.

Responses to Item 24 suggest that it is the researchers' responsibility to suggest how their research should be applied. This attitude is "politically correct"; there have been efforts over the years to get researchers to indicate the design implications of their work, but the results have so far not been very satisfying.

Item 29 put the matter plainly: Research should be applicable and useful beyond the production of more research. The response was resounding agreement, but what is disturbing is the strong disagreement position that, in effect, rejects research use.

Responses to Item 47 repeated those of Item 15; most respondents rejected the notion that research fails to contribute to design guidelines.

It is obvious that researchers genuinely believe that their work makes a contribution, but it is not clear as to what that contribution is. This cannot be known without some empirical investigation of information transmission from researcher to user.

Research Relevance.

7. The measurement *situation* must in large part *replicate* or *simulate* the *real world* conditions under which the human and the system ordinarily function. (A-43; D-16; N-3)
12. Criteria for the evaluation of research effectiveness *need not* include *relevance* of the results to solving or understanding real world problems. (A-11; D-42; N-8)
31. The *real world* operational environment is the only measurement venue in which research results can be completely *confirmed.* (A-33; D-23; N-4)

The themes of these items suggest that HF research results must be related to the real world in which systems and people normally perform their designed functions.

It would have been expected that this proposition would receive unanimous support (after all, who can be against the real world?), but it did not in Item 7. A sizable minority (16) rejected the concept. There may be several reasons for this rejection.

First, the nature of the measurement being performed may not require such replication. For example, anthropometric measurements are equally valid whether the measurements are taken in a laboratory or on a factory floor. Questions tied more closely to the human than to the human's technological (system) context, such as visual processes, for example, may require little or no simulation. The more strongly a behavioral function is affected by technology, the more that technology must be simulated in the measurement situation.

Another difficulty may be the respondents' inability to define the "real world." "Reality" is a philosophical concept whose definition is often idiosyncratic, varying with varying contexts. Researchers who are not system oriented may have had different notions of how to describe the real world. To others, the real world may appear to be uncontrolled and "messy," and therefore not measurable.

Item 12 suggested that relevance is not a criterion of research effectiveness. A majority rejected this point of view, but again a substantial minority (11) supported it and 7 respondents adopted a neutral position.

Item 31 suggested that only the real-world venue will permit complete confirmation of research results. A substantial minority (23) rejected this view.

The inability to define what is meant by the real world may be a source of some difficulty, because there are gradations of reality. For example, is a fixed, nonmotion automobile simulator an adequate representation of the real world of driving? In government system development research there are different levels of measurement (i.e., 6.1, 6.2, 6.3, etc.), a major differentiating factor being the degree of identity with real-world physical systems. This question rarely arises in conceptual research.

The lack of unanimity with regard to research relevance and reality is somewhat disturbing, because it suggests that some researchers perform their work without any reference to anything beyond their immediate measurement situation.

The Importance of Control.

18. Measurement without *control* (standardization of conditions) of the measurement situation leads to uninterpretable data. (A-46; D-15; N-1)
34. The researcher's need for *objectivity* and *control* over the measurement situation will largely dictate the *choice* of methodology. (A-44; D-12; N-6)

Control is an extremely important function of measurement, because until the measurement situation is structured by the researcher, that situation cannot be adequately measured.

Even if the researcher does nothing more than observe, that observation must be structured by asking: How shall I measure (performance criteria, measures, data collection procedures, etc.)? Structure that defines the objects and performances of interest equates to control.

Control is largely the result of researcher decisions. Individual researchers may perform all the measurement processes of Table 2.1 routinely, but most of the decisions they make are directed at achieving control. Structure in objective measurement is created by specifying the tasks subjects will perform and how they do these. In subjective performance (e.g., questionnaires and interviews), structure is provided by the questions asked and by the words used. Another form of control is exercised in the experiment by the selection of variables that form the basis of treatment conditions. Control is also exercised by the selection of subjects, because the subjects through their individual performances ultimately determine what the study will reveal. Setting up conditions to eliminate bias is another form of control, a negative control.

Item 18 suggested that without control, data would be uninterpretable. This proposition won general approval, although a substantial minority (15) disagreed. In Item 34, the need for control was presumed to exercise some effect on the type of methodology selected, such as the experiment, but again a significant minority (12) disagreed.

The reason for the disagreement with these items may result from the feeling that almost everything in the measurement situation is multidetermined, so that data clarity and method selection cannot be ascribed to one factor alone.

Measurement Venues.

22. If at all possible, all research should be performed in a *laboratory*, because this allows maximum *control* over the measurement situation. (A-2; D-56; N-3)
30. The *real world* operational environment is the only measurement venue in which research results can be completely *confirmed.* (A-33; D-23; N-4)

The reason for studying venue is to determine whether the venue, in the opinion of researchers, exerts a significant effect on performance. Objective evidence on this point is lacking, but researchers can be asked for their opinion.

Venue is not as unidimensional or as simple as it may seem. One dimension, particularly in the laboratory, is the amount of control the researcher can exert over the measurement. It is usually easier to control that measurement in the laboratory than it is in the other venues. Another factor is the extent to which the venue must resemble the operational environment (OE).

These factors ought to be primary in selection of a venue, except that one other factor often overshadows the others; this is the factor of *convenience* and where a particular type of subject is to be found. If a researcher is an academic, the availability of a classroom as a test environment and students as subjects may be more important than anything else. In a study of the aged, it is often necessary to make use of their customary living areas, retirement homes, senior citizen community centers, and so on.

One other factor is important: the nature of the task performance expected of subjects. If researchers are surveying preferences or opinions, as in the survey, the venue is unimportant, because the activity being studied is either independent of venue, or is under the subject's control and no one else's.

The preference for the laboratory in Item 22 was almost uniformly rejected. Item 31 suggested that the OE might be preferred, because it is a reference for all measurement behaviors, and thus complete confirmation or validation of study results can presumably be achieved only in the OE. There was a sharp split in responses, with a somewhat greater agreement with the proposition than disagreement. There are various ways of determining validity, some of which do not rely on comparisons with performance in the OE. Those who rejected Item 31 may have been thinking of these alternatives.

HF professionals are very much aware that there are other venues than the laboratory, and there are many occasions when other venues are more suitable. The response dichotomy for Item 31 may also reflect uncertainty about what the real world means to research. For those who disagreed with Item 31, the concept of a real world and its use as a referent for their work may have been obscure. This topic is probably not emphasized in university. The real world is so abstract a concept that the individual may be able to ignore it.

Research Effectiveness.

33. The primary *criteria* for evaluating research should be the adequacy of its *experimental design* and the precision of its statistics. (A-18; D-37; N-7)
38. The *evaluation* of research effectiveness is completely *subjective* and hence untrustworthy. (A-2; D-56; N-4)

If the research goal is to publish only papers of the highest quality, researchers must develop criteria for identifying that quality. Item 33 proposed that adequacy of experimental design and statistical precision should be criteria of quality research. However, such criteria would limit the evaluation to experiments only. Item 38 suggested that the evaluation of research effectiveness is completely subjective and therefore untrustworthy.

The concept of research effectiveness is hardly mentioned in textbooks and where it is considered, it relates only to the individual paper, not to the totality of research directed at solving a problem or a research program. The two items dealing with this topic did not, however, differentiate the evaluation of a single paper from evaluation of a program.

Respondents rejected almost two to one the notion that research effectiveness was no more than experimental design and statistics. On the other hand, almost one third of respondents agreed with the proposition, although their belief was relatively weak. Seven respondents opted for neutrality, which suggests some indecision.

The notion of evaluating research effectiveness is not, however, repugnant. Almost all respondents rejected Item 38, which suggested that research evaluation was too subjective to be trustworthy. Professionals believe that it is possible to say that some studies are good, others are less good, and still others are poor. Journal editors and their reviewers do this every day in evaluating papers for publication. The criteria used in these evaluations are entirely subjective (see Weimer, 1995, for a list of criteria), but respondents believe that it is possible to make such evaluations and to trust them.

Unfortunately, the evaluation of a single paper tells nothing about the effectiveness of a research program. It might be said that if individual papers are of high quality, then the combined papers, which comprise a program, must also be adequate. But does publication guarantee quality?

Arguably it is much more important to the discipline as a whole to evaluate a research program than a single paper.

The Research Community.

26. The *interests* of the research *community* as a whole, relative to a particular area of research, like workload, exert an important influence on the individual professional's selection of a research topic. (A-51; D-7; N-4)
40. What the HF research community as a whole is most *interested* in (as described in symposia and papers) has little impact on what the individual professional *decides* to study. (A-13; D-41; N-8)

If the research community is considered to be an intellectual system, then it seems logical that the community should influence the individual researcher, because individuals who are part of a system are by definition affected by that system. The community exerts its effect through its publication process, through its symposia, and through the university training it provides.

Responses to Item 26 confirmed almost unanimously the community influence; only seven respondents disagreed. In Item 40, the majority of respondents rejected the view that the community had no effect on research. However, some respondents were less affected by their peers. Thirteen respondents claimed the community had little effect, eight refused to vouchsafe an opinion. Responses to both items exhibited considerable variability in terms of how strongly respondents maintained their point of view; this suggests that the concept of a research community and its effects may be unclear to some professionals.

The professional community probably acts as a force for ensuring a certain uniformity among researchers, expressed in terms of topic and methodology selection. If the individual's research interests vary markedly from those of the community, they will be rejected. If, for example, a paper clearly demonstrating that telepathy could, under certain circumstances, predict human–system performance was submitted for publication, it would almost certainly be rejected on the ground that the topic was inappropriate for a HF journal.

The Store of Knowledge.

17. People talk about research contributing to the "*store of knowledge*," but no one really knows what this means. (A-18; D-35; N-9)
43. What we call "*knowledge*" is nothing more than *published* documents (articles, books, reports) describing research results. (A-9; D-50; N-4)

"Contributing to the store of knowledge" is often advanced as the ultimate rationale for performing research. The knowledge concept is also part of general science principles. The assumption is that knowledge in and of itself is useful, apart from any more mundane considerations.

Most (35) respondents to Item 17 did not agree with the proposition that no one knows what knowledge is; eighteen agreed. Nine of the respondents were indecisive.

Although 9 respondents agreed, most respondents (50) did not agree with the operational definition of knowledge in Item 43. Researchers may not know what knowledge is, but they do know (or at least think they know) that it is not confined to publications and to factual material. Again, the problem is a lack of an operational definition for knowledge, which permits disparate and individual definitions. It is probable that, as in the case of knowledge, many measurement concepts exist in the individual only as half formed implications and vague generalizations.

Speculation and Theory.

2. Hypothesis and speculation, which precede and follow data collection, are a *necessary* part of measurement. (A-51; D-9; N-3)
3. Theory is *not worth* much until it is tested empirically. (A-43; D-19; N-1)

Most respondents (51) to Item 2 agreed that speculation and theory are integral to measurement; only 9 rejected the concept. Most respondents (43) to Item 3 agreed that theory must be empirically confirmed, but a substantial minority rejected the notion, possibly on the ground that theory has uses unrelated to measurement and data.

The less than fortunate aspect of speculation is that it may be largely uncontrolled by data. Imagination (e.g., "what if . . .") is an essential element of speculation. If researchers do not label their speculations as such, then it is possible that some speculation may be mistaken for fact.

Miscellaneous.

11. The experiment is the *preferred* method of measurement in all situations where an experiment can be performed. (A-28; D-27; N-8)
37. Many HF professionals find measurement *objectives* like objectivity, validity, and precision overly *abstract*, somewhat unclear, and not very useful in performing actual measurements. (A-18; D-33; N-12)
39. All measurement depends on the *recognition* of a *problem*, whether physical (human/machine system) or conceptual (e.g., lack of needed information). (A-42; D-17; N-4)
51. The selection of a research topic determines all other research activities. (A-15; D-31; N-12)

Item 11 asked whether, given the opportunity, researchers would select the experiment over all other measurement methods. An approximately equal number of respondents agreed and disagreed with the proposition.

Item 37 suggested that perhaps abstract measurement objectives are not very useful in actual research. This proposition was rejected by half the respondents (33), although a substantial minority (18) supported it. Twelve respondents refused to commit themselves, suggesting a certain indecision about such concepts. Agreement with Item 37 may reflect the need for researchers to be perceived as scientifically "correct." It is also possible that certain very abstract concepts are retained in the CS because they are part of generally accepted scientific goals, but are rarely called on in actual research.

Such abstract objectives become more useful when they are turned into effectiveness criteria; it is possible to examine a study or series of studies and ask: Are its data objective, valid, precise, and so forth? Even then, because there are no operational definitions of these criteria, they cannot easily be employed in evaluating the study.

Item 39 suggested that all research was a response to a problem of one type or another. This seems a rather innocuous concept and most respondents (42) supported it, but again a substantial minority (17) did not agree. Obviously, research does not exist in a vacuum; something initiates it and that something is hypothesized to be a problem (empirical or conceptual) that creates a need for the research. Those who rejected the notion appealed to the job, which may require a particular type of research; or if an individual teaches in a university, the need to gain tenure would require research. To these respondents, these are not problems in a conventional sense.

Item 51 suggested that the selection of a research topic determines subsequent measurement functions. For example, if an aviation study is conducted, the fact that the system involved is an aircraft determines the kinds of tasks to be performed, the nature of the subjects to be used, measures to be recorded, and so on. The proposition was rejected by the majority (31) of respondents, although again a substantial minority (15) supported it. Twelve were neutral, suggesting indecision. It is likely that when the research is not related to or is only peripherally related to a particular system type, the nature of the research topic is far less determinative of subsequent research functions.

Skill.

19. Measurement effectiveness is significantly related to the individual personal *capabilities* of the researcher. (A-42; D-11; N-5)

It seems logical to assume that the more skilled the researchers, the better research they can produce. Most respondents (47) agreed, but as usual there was a minority opinion (11). Those who disagreed may have thought that job requirements, organizational factors, the constraints of contract research, and the environment in which researchers work have just as much influence as skill.

The concept "better or more effective" research may have little meaning for some researchers; perhaps all research is assumed to be effective.

CONCLUSIONS

From the results presented, the professional's CS is not monolithic. Each CS contains all the elements of Table 3.1, but the degree to which each individual believes in or acts on one element or the other varies widely (between elements and among individuals), and is highly dependent on the particular individual's experience and the measurement context in which each element is considered.

This variability is fostered by the lack of operational definition of the elements. Most CS elements are very abstract and function primarily as ideals that are honored more in their breach than in their observance.

CS parameters give rise to assumptions made by the individual that are situational; their values may change with the specific measurement problem. These assumptions are pragmatic; they are designed to enable the researcher to complete the measurement with least effort and (more important) in accord with the hypotheses that initiated the study. For example, the researchers know that their experimental conclusions are weakened by a small subject *N*, but they reason that this *N* does not fatally invalidate the study conclusions. This weakness is not publicized, in the expectation that readers will overlook the weakness. Or, inferences based on statistical evidence are recognized as highly speculative, because the statistical assumptions behind the conclusions are not satisfied. The professional may assume that the statistical assumptions are not that important and the raw data suggest the study hypotheses are indeed verified.

Many of these "errors" result from ignorance, particularly of statistical subtleties (see, e.g., Boldovici, Bessemer, & Bolton, 2002), but supporting the errors is the half-conscious assumption that such "errors" do not really cancel out the researcher's conclusions. In fact, even before the data are analyzed, professionals may suspect that they have an inadequate number of subjects, or may assume that variations in subject background are too minor to influence the results; or, that unequal *N*s in comparison groups do not confound the results, and so on. Professionals often make assumptions that scientific criteria, like relevance, validity, reliability, and so forth, are only goals and not "hard and fast" requirements. Customary measurement practices interact with and support these pragmatic assumptions.

The preceding also suggests that there is a built-in tension between the scientific goals of certain CS elements and customary measurement practices that are also part of the professionals' CS. Customary measurement practices are those habitual measurement procedures or strategies (always more concrete than scientific elements) that are performed by the individual as a matter of course, almost without conscious consideration. In any particular measurement situation, scientific goals are assumed to be either irrelevant or are opposed to the demands of the situation. Customary measurement practices, being very habituated, take over control of the measurement procedures. This last occurs in part because these goals (validity, etc.) are rarely in the forefront of the individual's consciousness and, being quite abstract, are less powerful than more concrete customary practices.

Each CS element is multidimensional, consisting of a core concept that is accompanied by a number of subordinate concepts representing alternative contingencies. It is these last that give the CS its indeterminacy. For example, validation is defined as a confirmation of research results, but it has subordinate concepts relating to different confirmation methods and venues. The more complex a concept, the more the concept is bound together with subconcepts, like alternative ways of performing a validation. Where the concepts are multidimensional, more than one factor influences its interpretation.

Most of the concepts in the professionals' CS are poorly defined, or not defined at all, which is another reason why respondents produced so many alternative interpretations of the concepts. Because of the lack of definition (in part because such concepts are not ordinarily taught in university), each concept must be defined anew by the researcher for each new measurement problem examined.

The more abstract level of the CS is probably just below the level of consciousness for most researchers; they rarely think of those concepts until their attention is forcefully called to them. In part, because of this, some survey respondents may have experienced difficulty and were indecisive in interpreting the concepts.

Concepts have attached to them a confidence factor, based possibly on the respondents' experiences. Individual respondents manifested a wide range of confidence values, indicated by where they checked on the agreement/disagreement scales (i.e., strong, moderate, weak, etc.).

The CS can be thought of as a behavioral resource like any other resource, such as physical strength or attention. The CS is tapped only when an actual measurement problem presents difficulties that involve concepts. That may be why the impact of the higher level CS elements has little influence on measurement practices.

It is the measurement problem that elicits the concept interpretation. It is hypothesized that a concept is activated when the actual measurement problem presents a difficulty requiring consideration of that concept. Respondents were aware of the importance of context, because some voluntarily reported that in one context they would respond one way, and in another context, in a different way.

There is a glaring gap between what researchers apparently believe in strongly as goals, like validation and prediction, and efforts to implement those goals. Respondents tend to blame external forces, such as funding agencies, for inability to accomplish such goals.

The researchers' CS is initially developed in university as background to their training in customary measurement practices (e.g., how to plan a study, how to select a data collection method, etc.). The original CS is later modified, strengthened, or weakened by the researchers' experience with actual research problems. These experiences lead to the development of pragmatic assumptions that relate the CS to measurement practices. These pragmatic assumptions mediate the effect of the higher order CS on practices, by in effect canceling out the higher order concepts when they are too difficult to apply to real problems.

The specific research problem to which the CS is applied is encapsulated in a specialty area that reduces the number of degrees of freedom the researcher has in attacking that problem. For example, researchers performing aviation research must organize the measurement in terms of aviation dynamics and the cockpit. This means that the selection of a research topic (which to some degree directs the further research process for that topic) is less determined by the needs (goals) of the discipline than by the needs of the individual specialty area in which the researcher works.

The major function of the CS appears to be to serve as an overall organizing framework within which the customary measurement practices can be fitted. The CS contains concepts that serve as criteria to evaluate and justify researcher decisions; that describe and, again, rationalize differences in measurement, such as basic and applied research; that enable the researcher to relate to higher order scientific principles like validity and objectivity; and that relate HF concepts to real-world factors that affect measurement. The CS contains goals like the purposes for which HF measurement is performed and the need for validation and prediction. It contains a rationale for failure to achieve those goals, such as the absence of required funding. Overall, what the CS does for researchers (whether or not this is a proper function for a CS) is to instill confidence that they have performed measurement practices correctly.

Assuming that the certain limitations of HF research stem from CS characteristics, what can be done to influence those characteristics? It is not easy to effect changes in a CS, but a first step is to begin to discuss the CS, and in particular to try to define operationally what each of its concepts means in relation to real-world measurement problems. Some very concrete steps can be taken by the universities; it may be too late to change traditional modes of thinking in those with many years of experience, but the universities might be able to open up conceptual horizons for students. The HF learned societies can help by providing a forum in which CS questions can be discussed.

Unmanaged Performance Measurement

Table 4.1 and Fig. 4.1 list the various objective and subjective measurement methods available to the HF professional. The first item on the list is something called unmanaged performance measurement (UPM). This chapter is devoted to UPM. It first examines the parameters that should be considered in any description of measurement methods. Chapter 5 explores the experiment and other objective methods and chapter 6 describes subjective methods.

MEASUREMENT PARAMETERS

Descriptive Parameters

The following are the parameters that influence the manner with which all the methods are performed and should therefore be included in their description:

1. The *objectivity* with which performance can be measured.
2. The *venue* in which performance is measured (i.e., the operational environment [OE], the laboratory, the simulator, the test site).
3. The degree to which the OE and/or the operational system must be *simulated.*
4. Whether or not subjects and variables are *manipulated.*
5. Whether the performance is recorded manually or *automatically.*
6. Whether performance is represented in physical form or *symbolically.*

TABLE 4.1
List of Measurement Methods

OBJECTIVE METHODS
Performance Measurement
1. Unprogrammed performance measurement
2. Experiment
3. Automated performance measurement
4. Simulated test situation
5. Analysis of archival data
6. Human performance predictive models

SUBJECTIVE METHODS
Observational
1. Observation
2. Inspection

Self-report
1. Interview
2. Debriefing
3. Questionnaire
4. Knowledge elicitation
5. Critical Incident Technique
6. Verbal protocol
7. Demonstration
8. Consensus measurement (e.g., focus group)

Judgmental
1. Psychophysical methods
2. Scaling methods
3. Comparisons (e.g., paired comparisons, comparisons of guidelines and standards with equipment drawings)
4. Decision methods

Application
1. Forensic investigations

Objectivity/Subjectivity. The first major parameter can be thought of as, but is really not, a dichotomy. In objective performance, the data can be recorded without the aid or expression of the subject whose performance is being recorded. The subject performs a task and that performance is recorded manually or automatically; the subject does not comment on the performance verbally, unless this is required by the task being performed.

In subjective performance, the data make use of the subject as the *medium* for expressing those data. For example, the subject may provide information in response to questioning, may describe his or her thought processes during task performance (in the form of a verbal protocol), may make a judgment by means of a scale, and so on. In subjective performance, the performance is the direct result of the subject's conscious communication.

There can be, of course, no complete dichotomy between objective and subjective performance. There is always human involvement in measure-

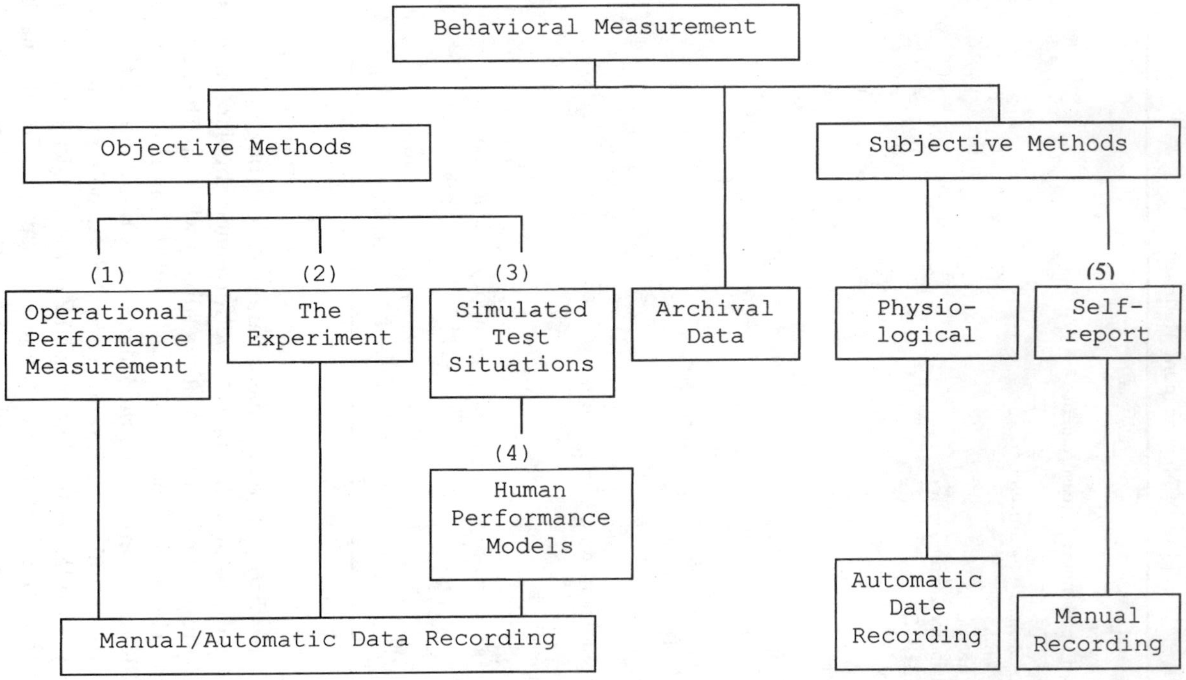

(1) Through observation
(2) Through segregation of variables
(3) Through simulation
(4) Through symbolic means
(5) Through self-consciousness

FIG. 4.1. The derivation of behavioral measurement methods.

ment, if only because it is a human who performs and a researcher who records and interprets the subject's performance. It is more correct to define objective performance as *minimal* human involvement in the expression of that performance.

Venue. Performance can be recorded in any environment. However, certain venues are more compatible with a particular method than are others. For example, the experiment is not easily performed in the operational environment (OE), because the manipulation of variables, which is a major characteristic of the experiment, is difficult to accomplish in the OE. On the other hand, when the system being studied is large and complex, such as performance in a factory, it is easier to study the system in its working environment. That is because large systems are difficult to simulate physically. The venue in which a problem is studied is often determined by the ease with which one can accommodate the problem to the venue.

Simulation. Simulation refers to the extent to which characteristics of the operational system and the OE must be reproduced in the measurement situation. Simulation can be somewhat simplistically defined in terms of similarity to the real world.

Simulation is not an all or nothing process; if certain characteristics of the operational system and the OE are considered to have only a minor effect on system performance, these can be ignored. For example, in a simulation of a warship navigation system, it is unnecessary to simulate the ship's galley, or the precise color of an equipment.

Researchers simulate because they wish performance data to be representative of the actual system and its personnel, and applicable to the OE in which humans and systems routinely function. The assumption is that, for certain problems and questions, the operational system and the OE have or have not an effect on human–system performance. For example, in psychophysical studies of human capacity, the OE has only a little effect on the expression of that capability. That is because these studies deal with psychophysical capabilities, like depth perception, which, because they are inherent in the human, are relatively immune to OE effects. So, it may be assumed that such studies can be validly performed in a laboratory.

If researchers conclude that operational characteristics may have an appreciable effect on performance, then they must logically perform the study in the OE, or, if this is not possible or desirable, they must replicate the characteristics of the system and its environment in the measurement situation.

The extent of simulation required may vary from none (rare in HF research), to partial, to complete simulation (as much as one can). Because there are no rules for deciding on this, it becomes a matter of hypothesis and judgment. Complete simulation is very expensive, so it becomes necessary to decide—in the light of the research and questions asked—how much simulation is necessary.

There are two types of simulation, *physical* (this is what HF professionals are most concerned with) and *behavioral.* The latter is a function of the former; it is assumed that if the physical simulation of the system and the OE is complete, then behavioral simulation (i.e., the responses made by test participants) will be the same as those of actual operating personnel.

No one can be sure of this; it is only an assumption, but if test subjects have the same characteristics as those operating the actual system, the assumption is likely to be true. More than this, it would be difficult to say, because no one has studied behavioral simulation.

At least four dimensions of physical simulation can be identified (Meister, 1998), as can be seen in Table 4.2, which lists the alternatives to complete simulation in decreasing order of similarity to the actual system and OE. The percent of papers published in the 1996 and 1997 HFES annual meeting *Proceedings,* as these relate to the four dimensions, is listed in parallel. These percentages suggest that in many research instances measurement situations are poorly representative of actual systems and their OEs.

Manipulation of Variables. Most measurement does not involve the manipulation of variables, although a review of the pages of behavioral journals might make it seem otherwise. In experimental manipulation, variables are manipulated in the form of treatment conditions, in which certain (independent) variables are contrasted with other conditions in which other variables (controlled) are held constant.

Method of Recording Data. At present, most data are probably recorded manually (using a researcher as the recording medium, with or without instrumentation such as a video camera recorder or audio tape recorder). Increasingly, however, data recording mechanisms are being built into the equipment being operated (automatic recording). Both manual and automatic recording have implications for data accuracy and interpretation.

Symbolic Data. Most behavioral performance is physical, that is, the recording of data describing actual humans performing in the measurement situation. With the advent of computer technology, it is now possible to develop and exercise human performance models that symbolically represent both the human and the system in which the human functions. Symbolic humans in the model can be activated to represent their performance in a mission operation. This permits the researcher to collect symbolic performances comparable to those of the actual human in an actual system. The result of exercising a symbolic performance is to make a prediction of what

TABLE 4.2
Simulation Dimensions and Percent of Papers Representative of Each Dimension

Percent	*Dimension*
	MEASUREMENT ENVIRONMENT
8.5%	(1) Actual OE
2.0%	(2) OE reproduced in film and/or sound
1.5%	(3) Test site
11.0%	(4) High fidelity full system simulator
12.5%	(5) Medium/low fidelity simulator; part-task simulator; mockup
46.0%	(6) Laboratory
16.0%	(7) Other (e.g., classroom, survey)
	MEASUREMENT UNIT SIZE
22.0%	(1) Full system operation
4.5%	(2) Subsystem operation
13.0%	(3) Task/workstation
4.0%	(4) Part task
6.0%	(5) Individual control/display operation
50.0%	(6) Nonoperational equipment or synthetic system only
	PERFORMANCE MEASURED
12.5%	(1) Mission
12.5%	(2) Function
17.5%	(3) Task
7.5%	(4) Subtask
24.0%	(5) Nonequipment-related task (e.g., pegboard)
25.0%	(6) Other (e.g., accident reports, environmental measures)
	TASK CHARACTERISTICS
14.5%	(1) Real-world activity
31.0%	(2) Simulation of real-world activity
6.5%	(3) Abstraction of real-world activity (e.g., computer graphics)
26.0%	(4) Nonequipment-related task
21.0%	(5) Nontask activity (e.g., body movements)

data actual physical performance will produce, once that physical performance is measured.

All of the aforementioned parameters interact, so that measurement methods can be applied in various ways. For example, objective performance recording can be followed by recording of subjective data; data in the same measurement situation can be recorded both manually and automatically. Each of the methods in Table 4.1 can be and often are combined in the same measurement situation.

Method Description

Each method in this and the subsequent chapters is discussed under the following headings in their abbreviated form (represented by the italicized words):

1. *Description* of the method. What is distinctive about the method?
2. *Purpose* of the method.
3. *Use situations* in which the method can be employed. Although all methods can be used in every measurement situation, each is useful primarily in one situation.
4. *Questions* that the method answers. This is not solely the purpose of the method (which is only a general statement). The specific questions addressed by the method are listed in Table 4.3.
5. *Assumptions* underlying the method. All methods make assumptions, which the investigator should be aware of.
6. *Procedures* required if the method is used. The procedure may impose certain constraints on the research and on the researcher.
7. *Advantages* of the method. Why should one select this method rather than another?
8. *Disadvantages* of the method. All methods have costs in terms of the effort, time, instrumentation, and subjects demanded by the method.
9. Measurement *outputs*. What data outputs, which are listed in Table 4.4, are provided by the method?
10. *Summary* impressions. This is a comparison of the positive and negative aspects of the method.

TABLE 4.3
Questions to Be Asked of the Measurement Situation

1. What research theme should be studied?
2. What specific questions does this theme ask?
3. What method(s) should be used to collect relevant data?
4. What requirements are imposed by the selected method?
5. In what venue will the study be performed and what implications does venue selection have?
6. Does the operational environment (OE) in which measurement functions and tasks are performed have any effects on those functions and tasks?
7. Does the nature of the functions/tasks being studied have any implications for subject selection and data recording?
8. How much of the OE must be simulated in the measurement situation?
9. Once data are collected, what do they mean?

TABLE 4.4
Measurement Outputs

The following is not in order of importance:

Terminal Outputs

1. Amount achieved (e.g., amount learned, number of products inspected)
2. Performance achievement (i.e., Does the human–machine system satisfy requirements?)
3. Performance assessment (e.g., How well does the human and system perform as related to requirements?)
4. Comparative assessment (i.e., Which is better, two configurations, two methods, two test procedures, etc.?)
5. Preference measurement (i.e., user preferences in relation to prototype design)
6. Technology assessment (e.g., What technology characteristics, like icons, produce the most effective performance?)
7. Attitudinal responses (e.g., How does the subject feel about the test, and how does the worker perceive the system?)
8. Problem resolution (i.e., Has system redesign corrected the problem?)
9. Capability determination (i.e., threshold limits of human capabilities, e.g., strength, visual resolution, motor speed)
10. Hypothesis verification (i.e., Do the hypothesized variable effects achieve statistical significance?)
11. Knowledge elicitation (e.g., demonstration by a subject of special skills and knowledge)

Intermediate Outputs

1. Frequency and type of outputs (e.g., errors, response times, accept/reject decisions as in inspection, decisions, actions; communications, subsystem interactions)
2. Physiological responses to stimuli
3. Environmental responses (e.g., performance responses to light and sound)

UNMANAGED PERFORMANCE MEASUREMENT

Description

Unmanaged performance measurement (UPM) is the simplest and most basic of the objective performance measurement methods. If, for example, and only as an illustration, two individuals sat at a window overlooking a street and observed the passage of pedestrians and automobiles for an hour (say, 3 P.M. to 4 P.M.) during the afternoon, and if they recorded how many people passed by, how many laughed or had serious expressions, how many automobiles passed by, what they would be doing would be UPM.

The major characteristic of UPM is in one respect negative: no observer physical contact with personnel being observed and hence no control or manipulation over what these people did while they were being observed. The reason for this is that any such contact inevitably distorts the performance being measured. Measurement operations that seek to collect data from the human subject influence the performance measurement. The recording of subject performance might involve instrumentation (binoculars to observe faces; a timer, if observers were interested in the duration subjects were performing; a video camera to record visually what subjects did; and a tape recorder to record what they said, if anything), but these do not distort performance as long as they are not noticed by the performer.

Any control observers would have over the measurement situation would be reflected only in their choice of the time period during which to observe, and their prior determination of what would be observed and recorded (e.g., frequency of groups and number of cars passing). Further control would be exercised by the observers' decision about how they would analyze the collected data after it was recorded. Or, to ensure reliability, investigators might decide to have two observers to cross-check what they saw; this would also represent a control action.

The lack of control (except for that exercised by the investigator on the previous observation conditions) distinguishes UPM from the experiment, in which the subject and the subject's performance are controlled (to some extent) by the experimenter. UPM is also distinguished from subjective data collection methods in that no subjects are asked questions or asked to report their status or feelings.

Considering the great frequency with which the published HF literature describes the experiment as the method of study, it might be supposed that UPM is very uncommon. It is uncommon in the published HF literature, but in reality it is far more common than the experiment in unpublished research. Besides being basic in the sense that all measurement begins with UPM (experimental and subjective measurement being mere expansions of UPM), it is also more common because everyone (not just HF researchers) utilizes UPM. A person who measures window dimensions before proceeding to purchase curtains, a carpenter who measures a space to be filled by a board, a motorist who visually evaluates an open space in which to park an automobile, and a drill sergeant who measures with his eyes the "dress" of a trainee formation are all engaging in UPM. A thousand other examples would clinch the point, but such a list would be tedious to the reader.

Since researchers in UPM are not permitted any contact with subjects or the conditions of their performance, UPM cannot study variables in the same way as an experiment. UPM does permit investigators to develop hypotheses about the causes of performance variations, but they cannot test these except indirectly, by examining how different facets of the subjects' performances coincide or differ, and by weighing the preponderance of the evidence (all of this mentally, of course). UPM can isolate what appears to it to be causal mechanisms, but cannot test them.

The reader may ask why, under these constraints, the researcher would prefer UPM to other methods. The answer is that there are certain measurement situations in which nothing can be done using other methods. If, for example, investigators are allowed (by higher authority or the owners of a system) no contact with or modification of the measurement situation as it functions with an operational system or its OE, then they may have no choice except to apply UPM.

The UPM is the method of choice, moreover, if the intent of the measurement is simply to gather data about the effectiveness of operator performance and the characteristics of that performance. The specific variables to be assigned to the data can, in most cases, be determined by prior observation. The reader should remember that one of the major purposes of HF research is to supply data that can be used in system design; in other words, simple performance measurement without any need to test causal mechanisms. Even if UPM is not used for this purpose, it can be.

An experiment under the previous circumstances would inevitably distort ordinary performance (because at least one experimental treatment condition would require some modification of normal subject performance). The most important use of UPM might be simply to collect data (e.g., the incidence of error or the number of personnel interactions). Under these circumstances, an experiment would defeat the data collection purpose of the study by producing unnatural subject performance. Useful data can be collected under other than UPM conditions, but these other methods, primarily the experiment, tend to distort the data outputs.

The noninterference requirement of UPM means that any data the method secures will be uninfluenced by the research process. In contrast to the experiment, which is inherently artificial, UPM does not distort anything, because it is, hopefully, unobtrusive. There are many situations in which the investigation must be unobtrusive, as for example, if data were being collected on what personnel in a surgical setting were doing.

The use of UPM does not exclude use of other measurement methods, but these other methods must be performed after the UPM measurement is completed. Thus, it is possible to interview UPM subjects after the UPM data collection period is over; and, if desired, an experiment could be performed following a UPM study.

The situations with which UPM are associated are often system-related. A number of system-related questions call for UPM. For example, if researchers wish to determine how well personnel in a system perform, they obviously should not influence the performance conditions. It is true that if they wish during system development to compare two possible configurations to discover which is better, they could perform a simple experiment, because a test comparison requires an experiment. Moreover, routine system performance may have inherent in that performance a number of contrasting conditions (e.g., night–day, men–women, different watch conditions aboard a ship) that would normally require an experiment if a researcher wished to study them formally. However, these may be compared in the UPM without conducting a special experiment.

One of the differences between UPM and the other methods is that UPM studies are not usually published in the general literature. That is because UPM cannot easily be utilized in the search for causal factors (a preoccupation of HF researchers) and because UPM is utilized most often in system-related studies, in which most researchers and funding agencies are not much interested.

Purpose

The purpose of UPM is to collect data describing the what, how, when, and where of a specified performance (*why* may require interviewing subjects, if this has not already been determined). That performance may deal with the operation of a system or equipment. An activity analysis (Meister, 1985) is an example of UPM. When performance criteria exist or can be developed, it is possible to determine how well subjects perform by comparing actual performance against the criterion. As part of this comparison, errors, personnel difficulties (to the extent that these are overt), and factors that seem to impact personnel performance (e.g., error-inducing features such as the inadequacy of procedural instructions or technical manuals) can be measured.

The overall purpose of UPM is to describe performance as completely as possible without manipulating that performance. This permits the illustration of obscurities in system-equipment performance. Many systems are so complex that it is not until the system is exercised under routine use conditions that procedural difficulties that must be remedied are discovered. An example is Beta testing of software by users (Sweetland, 1988).

Use Situations

UPM is utilized when the intent is to determine what a system will do or how operators perform some equipment-related task. UPM is often required in system development to discover whether a new system performs to design requirements. Indeed, this is usually a contractual requirement.

Military systems that are not asked to function continuously are exercised periodically to maintain personnel proficiency, try out new tactics, and learn more about system capability. All of these system design/use situations require UPM, because the system is exercised only to observe its performance. There are performance criteria to evaluate system performance, but to do so does not require an experiment.

The size of a system may be such that it is unfeasible (in terms of cost and time) to perform an experiment with it (even if one were able to secure authorization to perform an experiment). It is only when the system is decomposed to relatively molecular components like menus and icons that system-related experiments can be performed. When the measurement question to be answered relates to the system as a whole, only UPM can be applied to its measurement.

In UPM tests related to system development and use, the engineering concern is foremost. These tests will involve behavioral considerations only to the extent that the human factors aspects in the system are considered by management important to the functioning of the system. That is because managers assume that the engineering function subsumes the behavioral (i.e., if an equipment works as required, then obviously the personnel who serve the equipment also perform satisfactorily). Workload, operator stress, operator difficulty in using equipment, and so on, are ordinarily not sufficient motives for behavioral testing unless they obviously produce a decrement in equipment performance.

The general measurement methodology described in chapter 1 applies to UPM as much as to the experiment, with some variations. In addition to HF specialists having to learn details of system operation (if they are not already experts in the field), it is necessary for them to determine the specific purpose of the measurement and the specific questions to be answered; the measurement criteria and standards to be used in evaluating the adequacy of system performance; who is to act as subjects and how data are to be collected, using what instrumentation, interviews, observation, and so on; and the statistical analysis to be performed to answer the study questions. All of this is identical with the general process for performing experiments, except that in UPM no experimental design conditions are set up and subjects are not manipulated.

UPM can be applied to individual equipments (e.g., washing machines, desktop computers), as well as to systems. In evaluating an equipment, which is only a small part of a larger system, the equipment may be tested apart (both physically and in terms of venue) from the overall system, but if there is an intimate relation between the equipment and the system, then the researcher will have to take account of the possible effect on the equipment of the overall system. Of course, UPM makes use of subjective methods (e.g., interviews, scaling, etc.), but only as a follow-up to the UPM.

The use of UPM in evaluating total system effectiveness as part of its development and use has been emphasized, but there are other uses that can be valuable. Difficulties in subsystems may require attention. For example, a researcher might wish to determine the error rate in delivering medicine to patients in hospitals, or the error rate in inspection operations in manufacturing facilities. Or a problem such as the recent tire failure recall in Firestone and other companies may require an investigation of the behavioral factors in the manufacturing process.

Questions

There are questions which the investigator must ask of the UPM situation (see Table 4.3), but there are also questions that the methodology is particularly geared to answer and of which the investigator should be aware. UPM is particularly useful in answering the following general questions: *Who* is performing and *how* well? *What* are personnel doing? *How* do personnel do what they do? *When* are they doing these things? Why personnel are performing is part of the overall research context, which is usually available to the researcher before beginning a study. The result of answering the preceding questions is an empirical description in depth of system or equipment personnel.

The primary question UPM answers is: What is the performance status of the entity being measured (what is going on)? Where the entity is a system, this question becomes: Has the system completed the mission successfully? Are there any problems that need attending to? The criterion as to what constitutes successful mission completion may have been specified in system documentation, but often this criterion is something the researchers will have to determine on their own by analyzing the behavioral aspects of the mission.

There is probably a difference between the engineer's definition of mission success, which often has a single physical dimension (i.e., kill the submarine, fly 2000 miles, etc.), and what the HF specialist may consider adequate to describe operator performance. The specialist will have to keep the mission's terminal goal in mind, but the specialist's concern is with the relation between what system personnel did to influence terminal mission success and that success—or failure. The manager's success evaluation for a system is much more simplistic than the HF specialist's mission success evaluation. This, of course, makes the latter much more difficult to determine.

Moreover, there may be immediate and longer term mission goals, which must be considered. The primary mission of a destroyer in World War II may have been to "kill" a submarine, but if the destroyer merely forced the submarine to stay down, thus permitting a convoy to proceed, the mission had been accomplished, at least in part. The point is that the system may have multiple goals, some of which may be accomplished more readily than others, some of which may be more covert than others, so it is necessary for the researcher to keep these multiple goals in mind when evaluating success.

All these problems of criteria and standards are peculiar to system status research, because the system is supposed to achieve a goal, and therefore it must be determined how well the goal is achieved; otherwise nothing can be evaluated. Evaluation is forced on the system researcher because the goal exists. Determining system success or failure is merely part of the UPM system description. The physical equipments involved in the system have only functions to be performed; it is the operator who has the goal, to whose implementation the function contributes. When the equipment is not directly operated on by personnel, the equipment has no goal, only a function.

In purely conceptual studies (nonsystem studies or so-called basic research), investigators do not have this problem of criteria and standards. Absent a system, the subject in a conceptual study has a goal—to accomplish the measurement task—but this goal does not extend beyond the task. In fact, it is possible to distinguish two different types of HF studies: one in which the system and its functioning is foreground (as in UPM), and another in which the human is foreground, and the technology exists, but only as background.

Assumptions

The following discussion is oriented around the system because most UPM is conducted in support of the system, subsystems, or major equipments.

Assumption 1. If the entity being measured by UPM is a system, all subsystems that are essential for system functioning must be complete and ready to function fully as designed. Any discrepancies in the system, if not fixed, are sufficiently minor that it is reasonable to assume they will not significantly impact system functioning. Managers who are understandably eager to certify the first production model of a system may insist that an UPM be conducted, even if major equipment items are not operationally ready.

Assumption 2. The system mission, its procedures, its goals, and so on, are fully documented and available to those who conduct the UPM. A task analysis and other supporting documentation, such as operating manuals, should be available.

Assumption 3. Quantitative criteria of expected system performance are available. It is impossible to ascertain system performance until such criteria are specified. It is reasonable to assume that no system will be developed without such criteria having been specified; however, these criteria are likely to be engineering criteria.

There are, however, times when mission accomplishment is a matter of judgment on the part of higher authority (e.g., the CEO of a company). It is likely that performance criteria for personnel do not exist, except those implied in procedural documents. For routine proceduralized operations human performance criteria are likely to be dichotomous; the operator performs correctly or fails completely. For complex systems whose behavioral operations are largely cognitive, such simplistic human performance criteria are not very revealing. Investigators performing system studies may have to develop their own criteria and get management to agree to them.

If behavioral performance criteria are to be useful to the HF researcher, then they must be linked to system criteria (e.g., mission accomplishment). Expenditure of fuel, for example, is not a behaviorally relevant criterion, unless the pilot miscalculated the fuel requirement. Only criteria derived from mission performance are behaviorally useful. In most cases, terminal performance requirements (e.g., killing a submarine, flying to a destination within a reasonable time period) are available. The same may not be true of subsystems and sub-subsystems. For example, what is a reasonable time for detection of an underwater object and its classification as submarine or nonsubmarine? Criteria are implicit in and must be extracted from required performance. If, for example, Task X must be performed in a specified time, that time becomes the (or at least one of the) performance criteria for that task. A problem arises when the performance requirement is unclear. What is a reasonable time for sonar detection of an underwater object? The sonarman would say that this depends on a number of interacting factors, such as distance from the target, water temperature, and salinity. Criteria can be developed under these circumstances, but they will have to be elicited from experts, and even then they may appear arbitrary.

Assumption 4. The conditions under which the system functions are those for which the system was designed, and during the UPM these are not disrupted by external (nontest) circumstances. Whatever is being measured needs to be complete, and subjects must be representative of real-world operators. If these conditions are not met, then erroneous data will result. The UPM simply reports what it measures, and if what is being measured is in some respect atypical of the real system, then the resultant data will also be distorted.

Procedures

It is necessary that members of the team that will conduct the UPM become fully conversant with the system and equipment to be tested. Each member of the UPM team may be qualified in measurement procedures, but these must now be applied to the specific entity under test. It is therefore probable that a period of time will be devoted to learning about the object of measurement. This means intensive study of task analyses (if available) and operating manuals describing the general mission of the system and procedures involved in the major subsystems. This will provide the minimal knowledge required to conduct the HF aspects of the UPM. That knowledge should be supplemented by observations of ongoing activities, and by interviews with system personnel (to answer questions and provide demonstrations of special operations).

In contrast, researchers who perform a conceptual study do not have to develop this specialized learning, because experimenters who perform conceptual studies do not ordinarily deal with actual systems and equipment. However, a comparable learning phase should also be required of researchers in conceptual studies, because any stimuli they present are linked to the behavioral functions performed in responding to those stimuli. The conceptual researcher should be aware of the anticipated effects of these stimuli. This learning is performed by assimilating the research literature about the type of stimuli being presented.

For major UPM system tests, development of a test plan to describe in the greatest possible detail how the test will be conducted is highly desirable. Essential sections of the test plan will include: purpose of the test and the measurement questions to be answered; listing of subsystems to be included in the UPM; specification of the criteria for mission accomplishment; measurement functions to be performed by each member of the measurement team; specification of the data to be collected; procedures for data collection, including use of specific instrumentation; specific measures to be employed and with which subsystem; how much data to be recorded (if the system functions in cycles); the measurement methods to be used (e.g., observation, interviews, etc.); projected analysis of the data to be gathered. Such detailed planning should also be performed for conceptual experiments, but it is not likely that most experimenters do so.

The UPM procedures should be pretested both generally and in detail. The pretest is needed to discover "glitches" in test procedures (and there will always be discrepancies, no matter how much care is taken). If the pretest cannot for one reason or another be performed while the system is operating, then something like a walkthrough might be used to check out the data collection procedure. The pretest should be part of every research study.

Advantages

Because UPM deals with a system performing normally in its operational environment or a simulation of that environment, the question of the relevance and the validity of the measurement does not arise. Validity becomes a requirement only when a general principle, usually involving the effect of variables, is to be generalized to other systems and situations. UPM of systems does not ordinarily involve this, because there is no attempt to examine variables. If the system performs poorly in the UPM, there may well be an effort to determine why, which may involve attempting to trace causal factors; but this is far from the testing of hypothesized variable effects. The possibility of confounded variables, which Proctor and Van Zandt (1994) suggested is the cause of invalidity, does not exist, because in UPM there cannot be confounded variables, unless the researcher measures the wrong behaviors.

Disadvantages

If the reader believes that the purpose of research is to unearth variables and their effects, then obviously UPM is a weak methodology. UPM researchers can hypothesize variables and their causal effects, just as the experimenter can; it is merely that the UPM does not test these hypotheses.

Although the method is not inherently complex, its application to large, complex systems may make its measurements clumsy, especially if the measurement extends to subsystems. It may, therefore, be necessary to reduce the size and complexity of the system by eliminating less important subsystems and equipments from the system aspects being measured. This is not a disadvantage of the method, but its application to large systems.

Measurement Outputs

Those outputs of Table 4.3 that do not depend on subject, task, or situation manipulation (as in the experiment), or in which stimuli are under the researcher's control, can be secured from UPM. UPM outputs include those of a terminal nature (amount achieved, performance achievement, performance and technology assessment, problem resolution, and knowledge elicitation). UPM also supplies intermediate outputs (e.g., errors, response times, personnel interactions) because intermediate outputs are essentially only descriptive. UPM is essentially descriptive and (when appropriate criteria are provided) evaluative. Hence, except for the study of causal factors (as in treatment conditions), UPM generally provides (when performed correctly) almost as much information as the experiment.

CONCLUSIONS

UPM is indispensable if all investigators seek to determine is what the system or phenomenon is and how it performs. UPM is also necessary when endeavoring to measure large or complex systems, for which the experimental method is simply not feasible.

Chapter 5

The Experiment and Other Objective Methods

The experiment is most commonly used in conceptual research to determine the causal factors responsible for behavioral variations. It is also used in system-oriented research to compare alternative technologies and methods. The experiment is the most well known of objective measurement methods, although there are other ways of securing objective data, such as those described in the last part of the chapter: the simulated test of personnel capability, automatic performance recording, and analysis of archival data (from both epidemiological sources and from HF research).

THE EXPERIMENT

The Experiment as a Comparison

Most HF professionals think they know what an experiment is, because they have been repeatedly exposed to it. The essence of the experiment is *comparison* of conditions. The conditions being compared may be of different types. In most system-oriented research, the comparison involves different system characteristics (e.g., software icons, menus, or operating procedures). In most conceptual (so-called basic) research, the comparison is in terms of treatment conditions based on the hypothesized effect of variables.

A great deal is said about variables in this chapter. They can be defined, although inadequately, as mechanisms responsible for differences in human performance. Variables are selected by the experimenter for treatment, the treatment consisting of selecting subjects to receive certain stimuli that are varied on the basis of the conditions defined by the treatment condition.

It is possible to think, metaphorically and simplistically, of these conditions as receptacles or bins in which the selected variables (called "independent" or "experimental") are extracted from the totality of all the variables presumably influencing the performance to be measured. The selected variables are assigned to one measurement condition (the experimental group) that permits these selected variables to exercise a special, distinctive influence on the performance of that treatment group.

For example, if experimenters wish to study the influence of experience (the experimental variable) on task performance, they find and segregate a group of experts in that task in the independent treatment group. The term *group* refers to personnel exposed to the treatment. A control group consists of subjects who will not be exposed to the special experimental condition. In the experience example they are represented by subjects who are not experts (novices). The two groups (representing the independent and control variable) are then compared in terms of their performance of the same task. Other variables (so-called dependent variables) may be used as performance measures.

The previous description is bare bones. Most experimental studies involve more than one experimental variable, and stimuli representing these variables may be presented in many different ways in order to cancel out potential biasing effects. Whatever the experimental complexities are, however, the essential principle of the experiment is comparison.

Experimental Designs

Mathematicians have created many ways of making these comparisons. Most relate to the arrangement of subjects and the presentation to them of experimental stimuli. For example, to exclude the influence of individual subject differences and learning effects during the experiment (effects that may obscure performance differences) subjects may be exposed to the experimental stimuli in varying order, and may also serve as their own controls. This last is in order to eliminate the variance resulting from differences in subject capability, and from subject learning.

This chapter does not intend to explore the varieties of experimental designs that are available; the reader is probably aware of books and chapters of books that have been written about these designs. The author's intent in discussing the experiment is to focus on the concepts behind use of this methodology. However, the experiment has become so identified with its experimental designs that some mention of these must be made.

Experimental designs are simply various ways of describing variations in treatment conditions, the assignment of subjects to those conditions, the order of subject testing, and the way in which data are to be statistically analyzed. A mystique has developed about experimental designs; so close is their association with the experiment that they tend to overshadow the concepts behind the method.

The most common experimental designs, as Williges (1995) described them (and most of the following is based on his discussion), are: two-group designs involving comparison of one condition with another (e.g., one type of software icon with another), using a *t* test or one-way analysis of variance (ANOVA) to analyze the results. More commonly, there are multiple-group designs, when there are more than two conditions (e.g., number of eye fixations as a function of three object sizes). These, too, utilize ANOVA. Each type of design can be classified by the way subjects are assigned to treatment conditions, such as *between-subjects design* (each subject receives only one treatment condition in the design, and subjects are randomly assigned to various experimental cells in the design). The *within-subjects design* is one in which all subjects receive every treatment combination, but in varying order to cancel out learning effects. A *mixed-factor design* exists when some design factors are treated in a within-subjects manner, and others are treated in a between-subjects manner. Other variations in experimental design are the following: completely counterbalanced, Latin Square, hierarchical design, blocking design, fractional factorial, and central composite designs. These variations are mentioned only to illustrate the variety and complexity of these designs.

Readers who are interested in the often abstruse mathematical descriptions of these designs should refer first to Williges (1995) and to other authors he recommended: Box and Draper (1987), Box et al. (1978), Montgomery (1991), Meyers (1990), and Winer et al. (1991). These experimental designs can be criticized (although they rarely are). The following is taken from Simon (2001).

The major assumption underlying the experiment is that when certain factors are held constant—controlled variables—the result of varying the independent variables (those exposed to test) permits the correct effects of these manipulated variables to be demonstrated. Simon pointed out that even though the controlled variables are not tested, they can still influence the performance effects produced by manipulating the experimental variables. It can be shown statistically that independent variables are not immune from the effects of controlled variables; the latter can still modify the performance results of the experimental variables. Simon argued that "failing to include all critical factors in an experiment will distort the experimental results" (p. 3).

TABLE 5.1
The Traditional Experimental Design Process

1. Begin with one or more hypotheses.
2. Focus on reducing the variable error component.
3. Randomize in order to perform valid experiments.
4. Counterbalance to neutralize sequence effects.
5. Replicate for improved reliability and error estimates.
6. Hold variables constant that are not of immediate interest.
7. Prevent confounding of effects.
8. Limit the size of the experiment to what is considered "doable."
9. Use the test of statistical significance as the basic tool for analyzing the data.

Note. Modified from Simon (2001).

Simon also pointed out that the traditional way in which experiments are performed (see Table 5.1) "wastes resources that might better be used to obtain more valid and generalizable results" (p. 5). The reasoning and statistics behind his arguments are too lengthy to be reported in this chapter, but they deserve consideration. The traditional steps listed in Table 5.1 are presented in classrooms and textbooks as a series of mathematical procedures to be followed, usually in step-by-step fashion. However, each step has consequences that can seriously distort the data. Because most researchers are not mathematicians, they take these experimental design procedures at face value. This means that any inadequacies in these designs are rarely examined and recognized by researchers.

Why Is the Experiment Popular?

The researcher's fascination with underlying factors producing performance may be one important element in the popularity of the experiment.

There may be other factors, too: The physical sciences, which have long served as role models for behavioral measurement, emphasize the experiment as the only truly "scientific" way of securing data from the measurement situation. The university is particularly important as the promulgator of the experiment. Another possible factor alluded to earlier is the researcher's preference for control of the measurement situation; if the researcher is looking for control, then the experiment is certainly the preeminent way of exerting control by direct manipulation of variables.

The Representation of Variables

The author's concept of variables is that these are performance-shaping factors (like performance variations resulting from operating under either day or night conditions). They may also be considered overt *representations* of covert invisible mechanisms. These mechanisms presumably cause the measurable effects on performance. What lies behind the performance-shaping factors or mechanisms is difficult to say. In the performance differences resulting from the day–night comparison, the latter may or may not be the directly causative factor; there may be something inherent in or associated with day–night operations that produces the experimental performance differences.

The more obvious day or night situation may not specifically induce these differences. For example, night operations might be associated with visual changes that reduce the use of cones, hence lessen perceptual performance. Consequently, the causal mechanism may be unknown, but may be linked to a more obvious condition that can be identified and thus varied. It may only be possible to speculate about covert causal mechanisms, although it may be possible to manipulate the overt conditions linked to them.

For example, the experience variable used previously as an example is represented by "experts," even though the experience factor in the human is, itself, actually an invisible construct.

As is evident from the survey results reported in chapter 3, some HF professionals dislike the notion that variables are invisible mechanisms (if they are invisible, then they cannot be controlled); they would prefer to reify these mechanisms by identifying them with the more visible treatment conditions.

Of course, variables can be defined operationally in terms of their overt representations. However, these representations are not the same as the variables themselves. For example, experience can be defined as years of performing certain tasks, or as a certain degree of skill in performing those tasks. In the first case (years), researchers assume that years of exposure to a certain type of work creates a skill that is the representation of experience. In the second case (evidence of skill), they can measure performance and use the performance as the evidence of the skill. Or they can rely on the judgment of subject matter experts as to who or what represents skill. The point that must be emphasized is that the variable, experience, is not the same thing as years of practice or degree of skill; these latter are simply associated with experience. Researchers do not observe experience, only its effects, because experience is an invisible mechanism.

The variable, such as experience, is invisible, so how it is defined for experimental purposes assumes some significance. If these invisible mechanisms can be defined operationally, by selecting subjects exhibiting the presumed effects of the variables, it is possible to manipulate the variables by assigning subjects to treatment conditions. Then researchers can deduce how important these variables are by observing whether subject assignment produced the hypothesized/anticipated effects.

The actual experiment (more specifically, its data collection) is not a process of identifying variables. That has been done before any subject is assigned to a treatment condition. The experiment tests whether the researcher's hypothesis (the selection of certain variables as causal mechanisms for the performance) was "correct." Correctness is defined by rejecting the explanation of treatment differences as resulting from chance. Data collection and data analysis are simply the end points of a highly complex conceptual process.

The experimental process has the following functions: (a) the identification of variables; (b) the development of hypotheses about these variables; (c) the manipulation of the selected variables in the form of alternative treatment conditions; (d) the selection of the most appropriate experimental design; (e) the assignment of subjects to the treatment conditions in the design; and (f) the performance of the experiment (data collection and analysis). Except for the physical data collection process, all the functions require the experimenter to make crucial decisions, some of which are more important than the choice of an experimental design.

The Importance of Variables

Some professionals say that the experiment is conducted to discover how important the selected variables are, but these variables were already deemed important by virtue of their selection by the experimenter. Researchers do not knowingly select unimportant variables (as they see them) with which to experiment. The color and size of a workstation used in a test are not called out as variables, because researchers assume that these variables do not materially affect operator performance. The experimenters "bet" that the variables they have selected are the important ones and then run the experiment to confirm or disconfirm their importance. The significance of difference statistics (e.g., the ANOVA) do not tell the experimenter how important some variables are relative to other variables, but whether the resulting performance occurred as a result of chance or as a function of the experimental variations that were measured.

The Identification of Variables

This process of identifying variables and selecting them for testing on the basis of presumed importance is not peculiar to the experiment, and is to be found in all behavioral measurement methods. For example, if researchers ask data collectors to observe the performance of people at a cocktail party, they must tell these observers what to look for. This, in turn, requires the prior identification of critical variables (or at least their performance equivalents) assumed to be functioning in the performance being observed. Those being observed will manifest a variety of behaviors, but the researchers will not require the data collectors to observe all of them; this would be difficult and expensive in manpower, as well as leading to observer error. Rather, researchers specify certain behaviors that are of importance to them (e.g., the number of interpersonal contacts). The behaviors so identified are those that merely *represent* the variables in which the researcher is interested. (The number of cocktail party contacts may be a function of the number of drinks available or the celebrity status of attendees.) Likewise, the specification of performance measures also requires the researcher to identify what is important about the performance to be measured.

Variables are more or less abstract or molar. The confirmation of a highly abstract variable, such as "experience," does not add greatly to the store of knowledge, because it is already known that experience does affect performance. Confirmation of more concrete variables such as the icon characteristics that make them more recognizable (McDougall, Curry, & de Bruijn, 1998) is much more valuable, because researchers can put the information about these more concrete variables to work immediately.

Although other measurement methods, such as UPM, identify variables, only the experiment directly manipulates them. These other methods, like UPM, influence measurement of performance only by selecting a situation (e.g., the bridge of a ship) for observation or a function in which the variables affecting performance can be expected to exercise an effect. Subjective methods like a questionnaire can also influence variables directly, by means of the way in which questions are worded.

What the researchers using a subjective method ask about (e.g., frequency of communications or personal interactions) indicates the variables they are interested in, as does the way in which they ask these questions. The experiment cannot investigate variables in this way, by directly asking the subject about them. The experiment studies variables through treatment conditions that are more abstract than the actual performance to be measured. The experimental treatment manipulates variables and subject performance *indirectly*, by establishing the conditions under which the subject *must* perform. The treatment condition is therefore the essential element of the experimental measurement situation.

In the experiment, if the measurement situation has been correctly developed, then the subject's freedom of choice in responding (alternative possible response) is restricted. This is where control is exercised. The experimental task usually requires a precise action that must be produced by the subject. This is different from the subjective method in which subjects must think first about how they respond.

The Confirmation of Hypotheses

It is taken for granted that the experiment is a device to confirm or disconfirm the hypothesis that certain variables exercise an effect on performance.

However, why do researchers need the experiment to confirm or disconfirm the hypothesized variables? The obvious answer is that the researcher could be wrong in developing certain hypotheses about variables, and this can only be learned by testing. However, even if the hypothesis is proven correct by a large significance of differences statistic (e.g., .0001), what then?

One of the effects of the experiment is to increase confidence in the researcher and the research community that the selected variables are indeed important. The question is, what can be done with this increased confidence? Not much, unless the confirmed variables have associated with them behavioral data that can be used either to help develop new systems or to clarify how systems and people perform. The confirmation/disconfirmation process has, by itself, only limited utility, as reflected in an addition to the knowledge store, but not technological utility (e.g., aid to design).

Take the variable of experience in a particular skill, like silversmithing. The statement that experience is important in this skill (as it is in others) is obvious; and the experiment confirming this reveals nothing new. But if the experiment tells that silversmiths must have at least a year of practice (experience) in that skill before they can perform adequately, that is useful information. A year's supervised practice can be required before certifying an individual as a qualified silversmith. Or, suppose experimental results indicate that before training an expected average error rate (however defined) would be .01, as opposed to an error rate of .0001 after a year's experience. That information has technological utility, because the measured error rate can be used to check that the individual is indeed qualified. This is not a novel idea; the medieval guilds required apprentices to produce a "master piece" before being certified as a master.

What One Wants to Learn from the Experiment

Statistical confidence level data (e.g., hypothesis confirmed at the .05 level) have little value for aiding design. The actual performance data values (i.e., a mean of 3.8 for Condition A, a mean of 5.6 for Condition B) may be useful in design, suggesting that experienced personnel will perform at a level represented by 5.6 (whatever that means). Whether actual experimental data are commonly used in this way is unclear, but it is unlikely.

Thus, the experiment is designed to confirm general principles (e.g., experience is important) that anecdotal evidence cannot confirm. However, the confirmation emphasizes only that experience is important, which researchers already suspected.

It is possible that researchers develop their experimental hypotheses as much on the basis of accumulated personal experience as on what previous research has told them. Certainly experience is a factor in the interpretation of data to develop conclusions; whatever result is too widely different from what researchers have learned to expect will be viewed with suspicion. Whatever is too similar to experiential lessons will be considered obvious. Is it possible that one purpose of the experiment is to verify cumulative wisdom?

The preceding is an exaggeration, of course. There are many principles derived from experimentation for which there is no anecdotal evidence. Usually these principles are associated with quantitative values, for example, the field of view expected using only rods. In these cases, the performance data are more important than any hypothesis confirmation, although confirmation of the general principle is needed before the data can be believed.

Confirmation of conclusions that are largely qualitative may be good enough for the experimenter, but not for other research users who wish to apply the experimental findings.

Research Users

Assuming that experimental results have uses, there are several types of users. The first user type is the researcher who performed the study, and other researchers who are interested in the same line of study. Researchers should be very interested in the details of the study, because their own work will build, in part, on any previous research. However, this scrutiny does not consider any potential applications of the research to real-world systems.

Then there are other professionals (mostly HF design specialists) who want to use the experimental results to provide guidelines about how they should design some system aspect, or to answer questions that arise about how operators should perform with a particular system configuration.

Experimental results may, however, provide the design specialist with inappropriate data, because the experiment is inherently artificial; as was explained previously, actual system operations do not contrast conditions. This artificiality may not matter, when the experimental conditions contrasted are closely linked to a type of system or a system characteristic. If, for example, an experiment contrasts differences in software icons (McDougall et al., 1998), then the results can be directly applied to design issues. When, as in much conceptual experimentation, the experimental variables are only remotely linked to technological features, the amount of inference required for HF design specialists to make use of the study results may be more than they can supply. Consequently, many experimental results may not be germane to the interests of design personnel. Munger et al. (1962) found only 163 out of 2,000-plus research reports that met their criteria of useful material.

Experimental Subjects

The experiment is designed by the arrangement of subjects. Once these are placed in treatment conditions, the researcher assumes that the treatment condition largely eliminates or at least reduces the effects of individual differences.

Too little attention has been paid to how subjects of a measurement study interpret the measurement in which they participate, how much effort they exert, and so on, and how these factors affect their responses. There is a tendency to think of subjects as being more or less completely controlled by the measurement situation. In a sense this is understandable, because researchers consider their subjects not as individuals but merely as members of a class (subjects), and it is how the class, not the individual subject, performs that interests the experimenter. The experimenter would probably prefer to have all subjects clones of each other, so that individual differences variance could be eliminated. Is it reasonable, however, to treat humans as if they were rats?

Individual variance may be less important if subjects are faced with a proceduralized task, in which their freedom to respond is constrained by the necessity of performing to the procedural requirements. However, present research (e.g., situation awareness studies; Endsley, 1995) often involves cognition, and cognitive functions are much more influenced by subject idiosyncrasies, particularly intelligence and personality. The experimenter usually assumes that these characteristics have no significant impact on the experimental findings. Almost never does the experimenter take these characteristics into consideration and test subjects for them. To estimate the potential effect of subject idiosyncrasy (outside of statistical analysis), every study should be terminated by debriefing subjects, even if what experimenters get from the debriefing is only incidental information.

An equally common assumption is that the use of a large enough subject sample, and an appropriate experimental design, will cancel out the possible effects of individual subject differences (whatever these are). Inadequate sample size is sometimes used to explain a rejected hypothesis.

The experimenter may also feel that the more remote the experimental task is from an operational one, the less need there is for specially qualified subjects, and therefore the less need to be concerned about their characteristics. If training is required on a synthetic (artificial) task, experimenters provide the necessary training (although they may have only a vague idea of how much training for that task is required).

When operational systems are the focus of a measurement situation, the problem of subject selection is conceptually more manageable. The system and the operations to which the study findings will be applied determine subject selection. Experience in system tasks may be the most relevant selection factor. Not so when studying situations that are less technologically relevant. In such cases, researchers often assume that the behaviors required by the experimental tasks are those common to all adults, and so individual subject differences can be statistically canceled out. Are these subject variables unimportant? Not really, or there would be no effort to cancel them out statistically. Certain variables (e.g., age, gender, or required experience) may influence subject selection, but rarely intelligence or personality.

Magnifying the Effects of Variables

An additional effect of the experimental treatment condition is to magnify the effect of the variables selected for examination. The treatment creates a situation without so-called contaminating elements, because these last serve to obscure the primary effects in which the experimenter is interested. In real life, the notion of contamination is completely irrelevant, because all performance effects are "contaminated" by the normal mixture of variables, all exerting their influence at the same time.

In the experiment, however, the treatment condition acts in much the same way as does a microscope, by increasing the resolution of the lens viewing the object of interest. In the previous experience example, the novice group (the control group) cannot perform as well as the experimental expert group, because the former has no experts with whom the novices can consult to take advantage of their experience. In real life, novices often consult with experts, and indeed are expected to do so as part of their training. Because they cannot do so in the novice treatment group, the effect of their inexperience is magnified (i.e., they tend to make more errors). Sometimes subjects in control groups are withheld training, and this makes the comparison with those who have had training even more striking.

There is nothing wrong with this, because the treatment conditions are designed specifically to isolate and magnify variable effects. If they did not do this, it might not be possible to confirm the superior effect of the experimental variable (e.g., the .05 level of significance might be .10). However, in terms of applying and generalizing the experimental results to real-world situations, the adequacy of the application may be distorted by the magnification of the experimental differences.

Experimental Artificiality

That is why it is suggested that the experiment is basically an artificial mechanism. In real life, human–system behavior does not function in starkly contrasting ways (which is the basis on which to compare experimental and control groups). The removal or segregation of one or more variables from the mass, from all those variables that influence performance concurrently, is not something that occurs normally.

Do the independent (isolated) variables studied in the experiment influence performance in the same way the same variables function in real life, as part of a complex of variables? This may be the essence of the validation problem, and it applies to all measurement methods. The use environment as differentiated from the research environment may have an effect on the way in which variables function.

Research Application, Generalizability, and Utility

Researchers generally assume that experimental results will be utilizable in and therefore will generalize to the actual operations of the system or phenomenon being measured; otherwise, what would be the point of doing the experiment?

Researchers almost never worry about the generalizability of their experimental results; they merely assume that their study results will generalize. In part, this is because studies of the generalization of research findings are difficult to perform and are consequently almost nonexistent. It is possible that if experimental results do not generalize to actual real-world operations, it means that performance has perhaps been studied in the wrong way—for example, by selecting the wrong variables for testing. It is also possible that there is some application and generalization of test results, but not nearly as much as the experimenter assumes. It would be highly desirable to determine what and how much has generalized, but that requires operational testing and comparison of results with the original study. This is not likely to happen.

The generalization concept is rather vague. It implies an *extension* of results from the test situation to a reality, which is what researchers wish to understand more fully. But generalization to what?: to use of the results, in which case generalization is application; or to more detailed description of performance in the OE? If that is the only effect of generalization, then would it not be more effective to observe a performance directly in the OE? Operationally, generalization turns out to be application, because if there is no application of the research results, then generalization is essentially irrelevant. In fact, it is irrelevant, if the researcher always assumes generalization, and no one knows what and how much generalization there actually is.

It is possible that researchers really do not understand what is meant by the concepts of application and generalization, because they have only an inadequate notion of the real-world operations to which they expect their results to apply. This is obviously not true when the experiment involves an operational system or a highly realistic simulator, but for many conceptual studies linked only tangentially to real-world systems, the conditions researchers hope their results will apply to may be somewhat vague to them.

The assumption by researchers that their experimental results will automatically generalize and apply ignores fundamental questions underlying the assumption: What aspects of the results are being applied, and to what uses? Who is using the results? How does the application occur? And, how useful is the material applied?

Is it the confirmed importance of the experimental variable that is now applied? The recognition of the importance of experience as a variable can readily be carried over to another environment and applied to another set of phenomena. However, the experience variable is already recognized by almost everyone as being important, so the transmission of this conclusion to professionals other than the researchers adds little or nothing. Indeed, what would be most surprising and important is if a series of studies showed that experience was not important.

There is no doubt that knowledge is useful; for example, knowing what the field of view is under certain circumstances could be most important in designing an aircraft cockpit. However, not all facts/data (if these are knowledge) are useful and it is necessary to determine what the parameters of useful knowledge are, rather than saying, as some theorists say, all knowledge is always useful under any or all circumstances.

What else might be passed on (generalized) to a user population? To a researcher, certainly, the answer would be the experimental data in the study. To the casual reader of a report, this simply reinforces the belief in the importance of the variable (e.g., experience). What would be important in generalization would be any data that would transform the general principle into something more task related and concrete. If, for example, the general principle suggested that a particular skill reduced average error rate by .0001 for every 2 weeks of training, that could be very useful.

The paradigm adopted by the author for knowledge use is that the research user asks a question of that knowledge. The question is asked concretely, quantitatively, in terms of certain effects that the knowledge will provide. General principles (e.g., experience is important) have no real-world value. Application assumes that concrete effects will result from the application. For example, data about the capability of the human to perform certain functions, how the human responds to some technological aspect, or the effects of environmental factors (e.g., illumination, vibration) are very useful.

Utility, applicability, and generalizability also tie in with *predictability*. The usefulness of a piece of information should be directly correlated with the ability to use that information to predict the performance of the human, the system, or the interaction of the human with the system. There is, then, a litmus test for the usefulness of so-called knowledge; if it enables making a prediction, then the information is useful; if it cannot be used for a prediction, then it has little utility. The prediction capability is important, not only because science says that a discipline should be able to predict, but more importantly, because the ability to use knowledge to predict validates that knowledge. All knowledge is not the same; some knowledge is worth more than other knowledge.

The lack of systematic studies of research application is a serious omission, because until the process is investigated and utility is confirmed, the meaningfulness of the entire research process, except as other researchers use the research, or as it adds to the knowledge store (whatever that is), is only hypothetical. This, of course, flies in the face of the fundamental science assumption that all knowledge automatically has utility.

The utility of research in the physical sciences cannot be denied. However, physical science has the advantage that, in addition to its use of research to develop new theory, it makes use of that research to develop new technology. This is more difficult for behavioral science, because these researchers are only one of a number of contributors to system development. It is, therefore, necessary to show that HF research also contributes to technology in general or system development in particular. There is, however, little effort to study the use of behavioral data in system development.

In studies of the use of HF information (e.g., Burns & Vicente, 1996; Burns, Vicente, Christoffersen, & Pawlak, 1997), utility has been measured by requiring subjects to answer questions that can be answered only by making use of a body of specially selected information. Questions are created in such a way that they can be answered only by utilizing a specific reference text; questions that might be difficult to answer, because the answers may not be found in the reference text, are avoided. The utility of information is measured by determining whether subjects can find the answers in the selected reference material. This situation is hardly compatible with the real world in which, when questions are asked, the answers may or may not be found in an easily accessible body of information.

The problem of research utilization is, of course, not specific to experimentation; it applies just as much to research performed with other techniques. However, the experiment is the source from which HF system specialists find much, if not most, of the material they attempt to apply.

In any event, the question of research utility is one that requires measurement, because the question will persist, even if ignored. One of the difficulties to be overcome is that research on measurement is methodological or "process" research, and most published research is content or specialty oriented. Researchers may think of measurement research as peripheral to their main specialty content interest.

Purpose of the Experiment

The experiment has two purposes dictated by the two types of problems in which it is used:

1. System-oriented problems: The experimenter attempts to determine which of two objects or phenomena is more effective from a performance standpoint, or whether the system performs as required.
2. Conceptual problems: The researcher attempts to confirm or disconfirm previously developed hypotheses about causal-factor variables.

In both cases, the purpose of the work is not to develop and provide data to be used by others. That is assumed to be an automatic output of the experiment.

Use Situations

The situations in which the experiment is used derive from the nature of the problem to be solved. The experiment as a comparison device is useful in answering both system-oriented and conceptual questions.

System and Subject Simulation. The extent to which experimental findings can be applied depends in large part on the degree to which the measurement situation replicates an operational system and/or the operating environment to which the findings are to be applied. The more similar these are, the easier the application is (of course, this is only a logical hypothesis, because no relevant research has been done to confirm this). If researchers could perform an experimental study on the actual system and in the actual use environment, then the application of the findings would be automatic. Experiments performed on very realistic simulators of the aircraft type are almost automatically applicable, but there are many other studies in which specific systems and operating environments are not replicated. Conceptual studies, even those that have a technological context, aim to develop general principles whose validity (i.e., of the principles) depends on the extent to which they can be generalized. What about the applicability of their findings? This raises fundamental questions of what is to be simulated and to what degree.

Simulation also applies to subjects. Where the measurement situation replicates an operational system requiring a high degree of skill to operate (e.g., piloting an aircraft), it is necessary to use skilled personnel as subjects. Such people have credentials that have been authorized by experts.

If a researcher is not replicating a specific system or tasks, what then? The fact is that for experimental tasks, which are not operational ones and are specifically developed for the study (i.e., synthetic tasks), subject requirements are almost always unknown. However, most experiments using synthetic tasks ignore this aspect. Biologists who make use of rats as subjects, the bloodlines of which have been documented over many rat generations, know much more about their subjects than a HF investigator who selects subjects from those available (e.g., as with university students) or who perform for a fee.

What is to be done? The hypothesis is that intelligence and personality influence subject responses; thus experimental findings need to be related to these. This is a topic for measurement research. If the hypothesis is disproved, well and good, then researchers can proceed without worrying about these potential sources of variance. If the hypothesis is confirmed, it may be necessary to develop a special occupational category—people who are tested for repeated use as professional subjects, just as certain rats are professional subjects.

Questions Answered by the Experiment

In the UPM, the measurement is of naturally occurring objects and phenomena, and their variables are not segregated, because the experimenter is not interested in determining causal factors. (It is, of course, possible to study two different situations, using UPM, e.g., two cocktail parties, varying only in the amount of alcohol provided and then to contrast the resultant data, in neither of which the participants are manipulated). In conceptual studies, the comparison is created by the experimenter in order to demonstrate the effects of variables that the experimenter has manipulated.

The experiment seeks to answer two questions:

1. What does the comparison show, and how significant are the differences between two or more conditions?
2. Do the effects of the variables the experimenter has manipulated confirm the original hypotheses?

(It is interesting to note the questions the experiment does *not* answer: What are the characteristics of the entity or phenomenon being studied? What is its performance relative to some standards? Researchers are presumed to know all they need to know about the object of measurement, and in many, if not most, cases there is no standard to which performance is to be compared.)

A consequence of the questions asked is the amount of inference required of the experimenter. For Question 1, comparatively little inference is required, because the question is relatively simple: Are the differences between the situations being compared large enough to be significant in a statistical sense? In Question 2, not only must the differences between treatment conditions be considered, but inferences must be drawn about *how* the variables affected performance. This requires a higher order level of explanation. That the variables had an effect on performance can be inferred from the findings from any method; but inferences in the case of conceptual studies are more directly required, because if one is dealing with causal variables, *indirect* (hence inferential) effects are involved.

Assumptions Underlying the Experiment

The basic assumption underlying the experiment is that segregating variables and contrasting them reveals their effects independent of other factors; that is, when variables have been correctly manipulated into appropriate treatment conditions, the performance effects that occur stem from the variables represented by those treatment conditions and not from any other uncontrolled and confounding factors. Thus, if performance differences are found between day and night conditions, then the variables responsible for these differences are indeed related to the differences between day and night conditions.

Other assumptions held by researchers are that the experiment provides valid and generalizable data (hence there is no need to demonstrate validity and generalizability), and the measurement situation provides a reasonable representation of the experimental variables as these latter would actually influence performance in real life. Every methodology makes these assumptions, but the assumptions are particularly important to the experiment, because the experimental segregation of variables into treatment conditions is something not found in real life. Both assumptions can be questioned, but researchers never do, perhaps because their attention is concentrated on the functions involved in subject performance; and these experimental functions are assumed to be the same ones involved in operational performance.

Procedures Required by the Experiment

Experimentation follows the general list of research functions in Tables 1.1 and 5.1. It can also be illustrated graphically by Fig. 5.1, which is roughly based on one by Williges (1995). Each function in Fig. 5.1 decomposes into subfunctions with their own requirements.

The following are comments on the major categories of Fig. 5.1.

Defining the Problem. Being able to select a meaningful research topic requires a certain amount of expertise in and knowledge of the relevant literature in the selected specialty area. The way in which researchers become involved in a particular research area may or may not be a matter of choice; their job may force the choice, although they may have a preference for one specialty rather than another. Beyond that, the speculation and theory described in previous research reports, the determination of contradictions and obscurities in that previous research, may suggest a new study.

Researchers must make an explicit decision to use the experiment as the methodology for the new study; an experiment may be contraindicated if the task requires very covert performance (almost everything occurs in the operator's head) or the measurement situation, such as a system like General Motors, is so large and complex that variables are not easily manipulated overtly. If researchers are going to perform a conceptual study, then the study must be conceptualized in terms of variables that can be manipulated.

Test Planning. Variables must be identified and operationally defined during test planning before researchers can proceed further. The variables to be manipulated must be subdivided into independent, control, and dependent variables. This is followed by selection of an experimental design (discussed earlier).

The experimental hypotheses must be accommodated to the constraints of the experimental design selected. The use of the term *constraints* is deliberate, because researchers may find that the most appropriate experimental design prevents them from using a desired test procedure; the assumptions or limitations of the experimental design may not permit what researchers would like to do. Compromises must then be made.

Conduct of the Study. The development of an experimental procedure requires answering very practical questions to implement the study purposes. For example, what is the nature of the task/skill to be measured? Will it be an operational task or a synthetic one? In the latter case, the characteristics of the task must be specified.

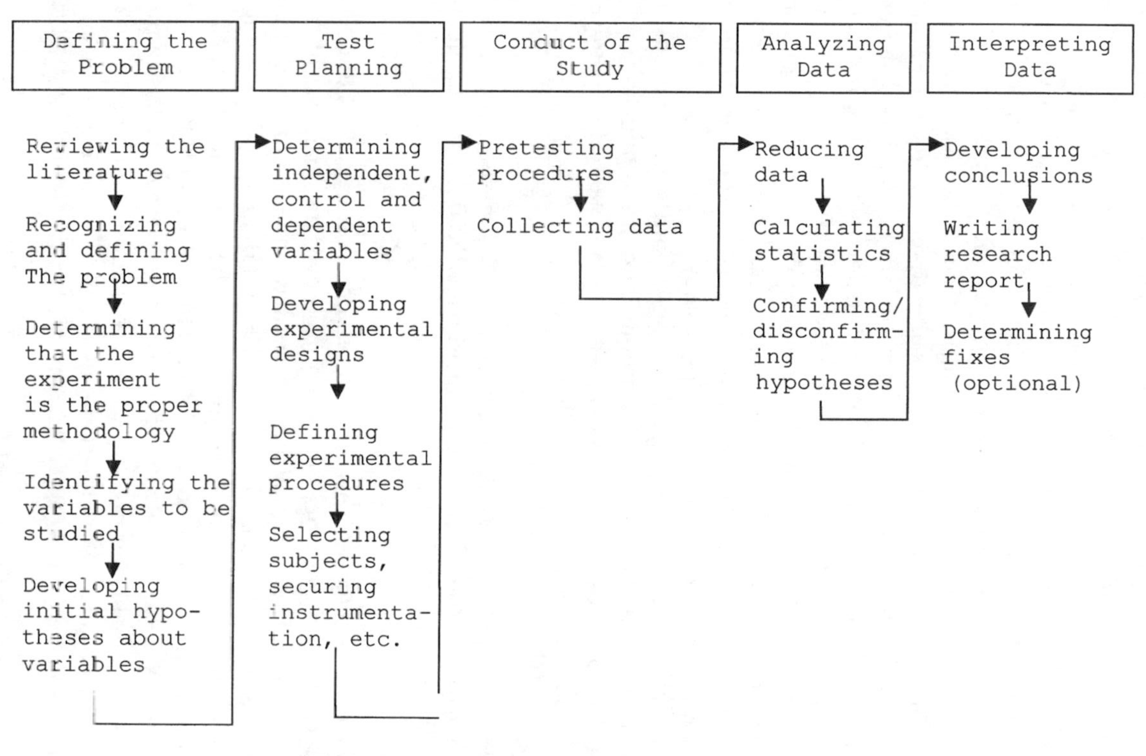

FIG. 5.1. Stages in the performance of an experiment.

The experimental design selected forces a particular procedure on the researcher (i.e., the details of how subjects will be arranged). Beyond that, there are very mundane questions of how stimuli are to be presented to subjects, for how long, how performance is to be measured, and so forth.

How can the experimental task, once developed by the researcher, be accommodated to the limitation of a maximum testing period, say, 1 hour? At this stage in test planning, the experimenter must develop a preliminary test procedure (e.g., 1 hour for training subjects in the task to be measured; 1 hour for subject performance; a retest after 2 weeks to measure skill retention). What kind of performance measures can be developed (a question that interacts with type of task)? What instrumentation will be required?

The experimental scenario requires the most effort in the test planning phase, because all the preceding work on identification of variables and development of hypotheses, selection of the desired experimental design, and determination of the kind of data to be collected must be fitted into the scenario, which should in all cases be pretested.

The actual collection of data may even afford some relief to the researcher, because when all the planning and pretesting has been performed, data collection often becomes a quasi-mechanical (repetitive) procedure, sometimes left to subordinates.

Data Analysis. Data analysis is largely "crunching numbers" (the data) through the computer and coming up with an ANOVA (or equivalent statistics).

Data Interpretation. This function is very closely tied to the "number crunching," but goes beyond that purpose. Data analysis will indicate whether or not the original hypotheses have been confirmed, but even when this question is answered, there are still questions of meaning to be resolved. Researchers are not "home free," because they have still to explain how the variables exercised their effect; it is not sufficient to report that they exercised an effect. Following this, the major task is to put it all into words, suitably sanitized, of course, so that the reader of the published paper has no idea of the angst the experimenter has experienced.

Advantages of the Experiment

If researchers are interested in testing variables, or making comparisons between conditions, then the experiment is the preferred method; no other method will do as well.

Disadvantages of the Experiment

There are, however, measurement situations for which the experiment may be contraindicated: (a) if the wish is merely to determine how people perform (to collect data); (b) if the object of investigation is a large, complex system from which it is difficult to extract and manipulate variables; or (c) if the functions studied are largely cognitive, so that overt, quantifiable responses are not easily available.

Another disadvantage is that the experiment may require some simulation of an operational environment in the development of the measurement situation, which raises a new crop of problems. There is always a certain artificiality in the experiment, because the operational situation to which its results must be referred does not lend itself to manipulation of variables.

It is necessary in many experimental situations to reduce the complexity of the measurement situation to the minimum required to secure a reasonable answer to the questions posed. This reduction can be achieved by controlling or eliminating all elements in the measurement situation other than those directly relevant to the test variables. For example, to study the effects of varying amounts of information on the performance of a command control team on an aircraft carrier bridge, ignore everything except the bridge and the communications coming in and going out. This reduces all communication inputs to those that are relevant to the measurement problem, which is to decide when the carrier should launch its aircraft in the face of a projected enemy threat.

Experimental Outputs

These are of two types: data derived directly from and descriptive of the human performance being measured; and conclusions from data analysis and their extrapolation in the form of speculation and theory.

Summary

All performance measurement methods involve complexities that are often not apparent to the researcher, but the concentration on variables makes the experiment more conceptually obscure than the other methods. The experiment, with its mathematical designs seems very clean cut, but when an experimenter begins to examine variables in detail, the experimental situation becomes more obscure.

OTHER OBJECTIVE METHODS

The measurement methods described in the following pages are all highly specialized. As a result, they are used only to supplement UPM and the experiment. The simulated test is used primarily to evaluate student capability: in public schools, as in an arithmetic test; in adult life, to measure automobile driving skill. The automated performance record is used to derive measures during and from equipment operations. The analysis of historical data is used to take advantage of epidemiological data, such as accident statistics, or human error reporting.

The Simulated Test

The simulated test (ST) is a test of personnel capability to perform required tasks. ST can be thought of as comparable to the tests given children in public school, except that in the context of HF such a test is system related and requires subjects to operate equipment. Examples are the periodic exercises performed by nuclear control room operators in a simulated control room, and the testing of sonar personnel performing simulated detection and tracking of underwater targets.

The ST can also be used for research. Many years ago, Meister (1971) studied how engineer-designers went about creating designs. The study was conducted in an office setting. Partial simulation was introduced by the presentation to subjects of a statement of design requirements (modeled on actual statements). The output was a design drawing. Following the drawing of the design, the subjects were asked questions about how they had gone about analyzing the requirements and developing the workstation design.

ST is somewhat like unmanaged performance measurement (UPM), because once the subjects are introduced to the measurement situation, there is no interference with their performance. On the other hand, ST is like the experiment, because its measurement situation is deliberately contrived by the researcher.

The following are characteristics of ST: (a) Measurements are made in a test environment and subjects know they are being tested for proficiency. The tasks they perform are simulations of operational ones. (b) The simulated test situation is a measurement of skill and the way personnel perform; it is not an experiment in which variables are manipulated. (c) The measurement situation, which often involves only a part of a total system and its mission, represents a problem that subjects must solve. (d) In measuring proficiency, there must be performance criteria and standards to determine performance quality.

Like UPM, the purpose of ST is to determine *how* and how *well* personnel perform a specific function or task. ST may be performed in any of the following: a training context, as in a school; in system development as a prototype test; or in a research setting, to perform research or as a method of gathering preliminary information that may lead to an experiment.

The major advantage of the ST is that it does not require the manipulation of variables, as in the experiment. This makes its measurement situation easier to construct. ST can be used to answer both empirical and conceptual questions.

On the other hand, if a ST is performed in order to determine how well personnel perform, it is first necessary to have performance criteria and a standard for adequate performance. This may require the researcher to develop those criteria and standards, if they do not already exist.

The ST is any objective measurement situation that does not involve experimental comparisons of variables. It differs from the UPM because it need not be performed in the OE using actual equipment. Consequently, the range of its application is very wide.

Automatic Performance Recording (APR)

One of the benefits of computer technology is that it now permits having the machine being operated by personnel record the latter's actions. This allows for automatic recording of what would previously have had to be recorded manually. APR can thus be thought of as a somewhat special form of instrumentation. It is used operationally; an example is the flight recorder aboard commercial aircraft, and it is often part of sophisticated simulators. APR is something different from common instruments, such as video or audio recording. The reason is that during its use, it does not require any human initiation, because it is directly tied into the equipment's internal mechanisms. This has consequences for the interpretation of APR data.

Any automatic means of recording data necessarily reduces those data to molecular mechanisms that the machine can implement. The machine cannot register operator intentions; it can record only subject responses. This means that researchers must work backward to infer the operator's purpose in order to understand what the machine recording means. This is true also of human-controlled instrumentation, but to a lesser extent in the latter. More inference is required of APR as to the meaning of its data, because the amount of human interaction is less.

The data interpreter must take into consideration the purpose of the action represented by molecular responses like computer keystrokes. That is because the machine and its operations express *physically* (but only indirectly) the operator's purpose in performing the recorded action (a keystroke is only a keystroke). The machine reveals what the operator intended, but not necessarily clearly. The machine implements the operator's intention—for example, a pilot wishes to land, and so pushes the control stick forward. That motion is what is automatically recorded, but the meaning of that motion is distinct from the motion itself. The intention behind the motion must be inferred.

Knowledge of purpose is essential in the interpretation of automated data, because the machine was developed only to implement a function, and the operator uses the machine as merely the means of implementing that same function.

The gap between the molecular physical representation of a higher order (perceptual/cognitive) action is found in all objective performance measurement, but is somewhat smaller in proceduralized or relatively simple cognitive operations like adding two figures together. The gap is greater and requires more intensive analysis, when the molecular action is recorded without human intervention and the operator functions performed are more complex. Adding two numbers is a highly proceduralized operation; deciding when to launch a missile attack by inputting commands via a keyboard is more complex by an order of magnitude. The inference from the machine action to the behavioral action that commanded it is easier when the latter is a motor action like controlling an aircraft landing gear.

APR does not make the process of interpreting human performance data easier for the analyst; it may even make it more difficult by increasing the amount of raw data that requires basic psychological operations—identification, classification, comparison, and inference. Where large volumes of molecular data are recorded, they are probably not all significant for the interpretation; it may be necessary to separate the wheat from the chaff. For example, the aviation flight recorder will record long intervals of level flight, when the analyst is perhaps interested only in the point at which the pilot decides to descend for landing.

Just as the design of the human–machine interface requires a choice of what is to be controlled by and displayed to the operator, so the design of the APR device requires choices as to what is to be recorded.

These choices require the APR designer to develop hypotheses about the information the analyst would wish to have in examining the machine record.

APR can be incorporated into any equipment, wherever that equipment is utilized, in research as well as in an operational device. All behaviors can be recorded, provided these are mediated by mechanical means.

The following are assumptions underlying APR: The machine actions recorded by APR reflect higher order, purposeful activity. Operator intent can be correctly inferred from APR data. And, APR data can be used in exactly the same way that manually recorded data can be used, provided the operator's intent can be correctly inferred.

The first assumption applies to all data in all measurement situations. Any performance measure is more molecular than the entire operator action it purports to describe. Therefore, it must be assumed that it reflects what the subject intended to do. The more complex the human operation, the greater the leap of faith needed to accept the validity of the performance measure.

An example is eye movement recording as evidence of cortical and, hence, cognitive function (Bedny & Meister, 1996). Eye movement data, as these relate to observation of graphic advertisements, have some relevance, because the duration of eye fixation as related to certain stimuli (e.g., photographs) should logically suggest interest in these photographs. However, duration of eye fixation says nothing about the nature of the subject's interest in those photographs, or the cognitive activity associated with the eye fixations.

Researchers analyzing APR data must keep in mind the questions they wish to answer. That is because APR records much more data than researchers are interested in. Measurement questions serve to classify data elements into those that are and are not relevant to these questions.

The difference between relevant and irrelevant data are found in all measurement situations and with all techniques. Observational data, for example, contain much irrelevant material, but the researcher creates mechanisms to eliminate the irrelevant material (e.g., a checklist of things to observe).

Where the measures to be recorded are many and quite molecular, APR can be quite useful. The human observer is not geared to record great masses of highly molecular, repetitive actions with manually operated instrumentation. At the same time, much of the information flow resulting from APR may be useless in a particular study, simply because the machine cannot discriminate between what is relevant and what is not for a particular set of questions.

Analysis of Archival Data

Archival data are *collections* of data gathered specifically to describe a particular topic. There are two types of such data, one with perhaps only incidental interest for the HF researcher and a second that is built into the structure of the HF discipline.

The first type is *operational statistics*. Its special characteristic is that it can be manipulated to discover relations that are inherent, although not fully expressed, in the data. A second type of archival data is found in *HF research reports*; it consists of all the performance data produced by HF researchers and collected in bound journals resting on library shelves.

Operational Statistics. Many agencies, usually governmental, or government related, require the collection of data related to their functioning. Types of data recorded include: accidents; census material; epidemiological data, such as birth and death statistics; hospital-related statistics, such as mortality and recovery rates in surgery; crime statistics; equipment maintenance records; and failure reports.

Although the original intent for which most of this operational data were collected was not for behavioral purposes, those data that are human related may be of interest to the HF professional, particularly those related to accident and safety issues. For example, in the 1950s, as part of continuing studies to predict human performance, HF investigators collected statistics in the form of equipment failure reports to determine how many failures were caused by human error (see Meister, 1999).

Because the data were collected for purposes other than those of the HF researcher, it is very likely that the conditions under which the data were collected were less than optimal for the purpose of behavioral data analysis. These data (e.g., accident statistics) are themselves summaries and reclassifications of the original data (e.g., traffic reports), so that what is available is "second generation" data.

Because the data may be incomplete, only partially relevant, and the conditions of data collection only partially known, the analysis may require considerably more extrapolation and speculation than would be needed with experimental data.

The purpose of analyzing operational statistics is to explore potential relations implicit in the data (e.g., traffic accidents are more likely to occur under low illumination conditions).

The use of operational statistics is likely to be quite circumscribed for the HF investigator because the original measurements were almost always made for purposes other than those of the HF researcher.

Anyone attempting to use such data must ask what relations of interest to the HF researcher are implicit in the original data. The specific questions the researcher can ask are determined by the classification categories imposed on the data by its original collector.

The basic assumptions underlying the use of operational statistics are that the data reflect an underlying reality and that reality can be discovered by manipulating the data. In fact, all data, no matter where they come from, make those assumptions; if those assumptions could not be made, then the data could not be used. In the experiment, the HF researcher is involved in that underlying reality by planning the research and collecting the test data, but this is not true of statistical data. If the statistical data can be secured, then the investigator will, as in the other measurement methods, examine the data in the light of hypotheses developed about potential relations in the data.

HF Archival Data. These data are found in collections of published journal articles subdivided into specialty subtopics like aging, computers, or aerospace.

HF research data comprise the formal knowledge of HF (the informal knowledge is in the minds of professionals). In their original state, as collections of individual papers, they are not interrelated; performance data in one study are not combined with other studies, except in historical references to past research in individual papers.

There have been a few attempts in the past to combine these data. For example, a HF team (Munger et al., 1962) published what they called the "Data Store," which consisted of error rates describing how accurately (and in what length of time) operators could read displays and manipulate controls. Data were collected, using a taxonomy of controls and displays, from over 2,000 journal articles. Based on certain criteria, 163 papers were accepted to serve as the data sample. Data extracted from these papers were combined and reformatted to develop probabilistic error rates to predict the accuracy of personnel performing simple perceptual/motor functions involving controls and displays. See Meister (1999) for a more comprehensive description of what was done.

The disabilities of operational statistics are not found in HF data. Almost all the data were gathered from rigidly controlled experiments whose conditions are described fairly completely in the published papers.

The principal value of HF data is that, when combined, they will permit the quantitative prediction of human performance in relation to various aspects of technology (primarily the human–machine system interface and corresponding operating procedures). Unfortunately, the development of predictive indices in the form of an integrated archive has rarely played a large role in the mainstream of HF research. The result is that, although the conclusions of past HF studies are referred to in individual research papers, the data in all these papers are never collected, organized, reformatted, and applied for predictive purposes.

HF archival data, if used for prediction purposes, require a certain amount of extrapolation and deduction, but the major process is the extraction of the quantitative data and their combination with other data, using a common metric, which is usually a human error probability (HEP) that equals:

$$\frac{\text{number of operations in which one or more errors occurred}}{\text{total number of operations performed}}$$

The combination of HF data for prediction purposes is constrained by the differences in the measurement situations from which the data were derived. In consequence, it is necessary to develop a number of different taxonomies centering on equipment, human, and human–equipment interactions in order to subsume differences in measurement conditions under more broadly descriptive categories.

For HF data analysis, the following questions should be asked: (a) Is there enough information provided in the research reports about the measurement situation, so that the human functions performed and the equipment characteristics involved can be identified? (b) Are the performance data reported in the form of errors, or some other suitable metric, or can they be reformatted to such a metric? The experience that Munger et al. (1962) had with their research reports (163 usable out of 2,000-plus reports) suggests that many reports lack necessary detail.

Assumptions Underlying the Analysis. One assumption underlying the analysis of HF archival data is that the differences between individual study situations are small enough that they can be combined in accordance with rules specified by the taxonomy used.

Another assumption is that the common metric of the error rate, which is the one generally used in predictive methodologies, can actually predict performance validly.

Procedures. The procedures involved in developing human performance data for prediction purposes are much more complex and require a much longer time period than those used for operational statistics, or indeed for data analysis in any single experiment. These procedures have been explained in more detail in Meister (1999), but the following are the major processes:

1. The development of a number of taxonomies to classify HF data in terms of three categories: human function, equipment function, human–equipment interaction. Presently available taxonomies (e.g., Fleishman & Quaintance, 1984) are excellent starting points, but require expansion.
2. A review must be undertaken of all past research within a specified time period, say 10 years, to extract the performance data from each research report by applying the taxonomies in Process 1. In addition to these taxonomies, it is necessary to develop and apply criteria for the acceptance of data (e.g., adequacy of background information, absence of confounding factors, up-to-date technology, similarity to the operational situations to which researchers wish to predict, ability to transform data into a common measurement metric).

Prediction requires use of a combinatorial model of how individual human–equipment values can be combined so that it is possible to predict more molar behaviors (e.g., functions and missions). A number of combinatorial models exist (see Swain, 1989). The model most widely used is THERP (Swain & Guttman, 1983).

The advantages of being able to predict quantitatively the human performance to be expected of personnel interacting with equipment should be obvious.

Disadvantages. There are a few disadvantages to the methodology. If the human performance data are collected from published experimental reports, it is possible that the artificiality of the experiment might tend to distort the prediction. It has been pointed out that experiments are designed to elicit subject errors so that differences between experimental and control conditions are magnified and, hence, easier to see. The increased number of errors may skew the prediction, so that it may be necessary to apply some sort of quantitative correction.

Beyond that, the combination of data from many studies requires a sustained effort to be carried out on a continuing basis; predictions based on prior data must be continuously upgraded by combination with new data.

Chapter 6

Subjective and Related Methods

Logically, an attempt should be made to define what is meant by subjective measurement methods, but there is an even more fundamental question: Why be interested in subjective measurement? Considering what HF professionals have been taught about the deficiencies of subjective data (the reliance on unreliable memory, the subject's unconscious biases, the subject's difficulty in assessing his/her internal condition), and the preferred status of objective methods (supposedly untainted by human agency), why bother about subjective data?

There are a number of answers to this question. First, people's own experiences suggest that private feelings and attitudes often influence what they do publicly, and the latter cannot be fully understood without learning about the former. There are some conditions about which researchers can learn only by asking subjects what they have been thinking, feeling, or deciding. This is the situation when the activity performed is almost entirely cognitive (and, hence, not amenable to instrumentation), or when the wish is to learn the subjects' perception of their internal status. For example, how can experimenters determine how much stress or workload subjects are experiencing without asking them?

Moreover, subjective perception, attitudes, emotions, decision factors, and so on represent a domain comparable to and functioning concurrently with that of objective performance, but different in kind. To ignore this domain is to "write off" the entire emotional/cognitive infrastructure that supports overt performance. Only the most confirmed behaviorist would claim that internal phenomena are of no interest to HF. The fact that the human, in operating a system, is at the same time thinking and feeling suggests the possibility that these internal functions can influence objective performance, and therefore it is necessary to investigate this possibility.

In addition, as pointed out in earlier chapters, one of the reasons for HF measurement is to determine the effect of technology on the human. That effect is experienced internally (e.g., stress) as well as externally.

Reluctance to study internal behavioral phenomena correlated with objective phenomena may occasionally result from the possibility that the former may sometimes conflict with the latter, in which case how is it possible to make sense of either? An objective situation may suggest high workload in terms of, say, a high communications requirement (messages coming in rapidly and overlapping each other), but when operators are asked if they feel stressed, they respond, "No."

The consideration of subjective phenomena as these relate to objective performance represents a form of the classic philosophic mind–body problem and, therefore, requires empirical investigation.

Moreover, many of the performance measures in HF measurement (see Gawron, 2000) are based solely on subjective indices. An example is the Cooper–Harper (1969) Scale, which measures aircraft flying qualities in terms of subjective impressions.

In conclusion, even if researchers wished to ignore subjective phenomena and subjective measurement methods, it is impossible to do so.

Which brings the discussion back to the original problem: How can subjective measurement methods be defined?

THE DEFINITION OF SUBJECTIVE MEASUREMENT

A subjective method obviously requires the participation of a human in the measurement process. However, that definition is too gross, because in these terms all measurement is subjective; a human is always involved in measurement, either as the subject of the measurement, or in its planning, data collection, or interpretation.

Subjective measurement might be defined as measurement that seeks to discover the subject's own internal status (e.g., symptoms, perceptions, concepts, attitudes). However, the subject also reports about external "objective" phenomena, as a witness; this definition is also incomplete.

The complexity of the definition problem is suggested by the various elements of subjective measurement:

1. *Who* is involved: the human who expresses a report; the human as observer, receiver, and recorder of that report.
2. *What* is recorded: both internal and external objects, events, conditions, phenomena.

3. The condition that *stimulates* the subjective reports: The latter may be self-induced, or it may be elicited by others (e.g., the researcher acting as interviewer) or by the technology itself (e.g., equipments that require a report).

4. The *purpose* of the subjective measurement: to determine the subject's perception of internal and external conditions, and to correlate these with objectively recorded performance.

5. The *time* status of the subjective report, which may describe present, past, or even future events. For example, the critical incident technique (Flanagan, 1954) describes past events; the measurement of situation awareness (Endsley, 1995) involves present events; the specification of an intention to perform an action related to technology (e.g., I intend to test this circuit) refers to the future.

It may appear to some readers that objective performance is goal directed, because it stems from the requirement to perform a mission or a task, whereas subjective performance, which parallels objective actions, is not. This is not quite correct. Although some internal status (e.g., workload, stress, fatigue) is not in itself goal directed, because it is merely correlated with or results from external actions, the perception of goal direction by the subject is entirely internal, but still interacts with and influences externally goal-directed performance. For example, if someone throws a switch, that person must first have the intention to throw that switch. Internal goal direction is important: If the individual forgets to throw the switch (a purely internal condition), the efficiency of a mission may be jeopardized.

Moreover, when subjects are asked to observe and report their internal status, the observation itself assumes a goal derived from the request: to report correctly, making the proper discriminations, and so on. This is, of course, a secondary, nonmission-directed goal. (Something, which has not previously been discussed, is that major goals are directed by a mission; secondary goals may be task directed, or derive their goal direction from simple association with primary goals.)

Perhaps the difference between objective and subjective is the *modality* used to make the recorded report. Objective performance is often reported by instrumentation (e.g., a flight recorder), and in subjective performance the modality is the human. However, this is not quite correct either, because humans can report physical actions without instrumentation, and automatic recording of external ("objective") events must be interpreted by a human.

Perhaps the *locus* of what is reported is what makes the difference. Maybe whatever is external to the human is objective and whatever is internal is subjective. However, the external event must be either reported by an observer or, even if recorded automatically, must be interpreted by a human, so whatever is external is therefore eventually internal and subjective.

Another possibility is that whatever can be understood immediately, unequivocally, is objective, and everything else is subjective (e.g., it is necessary to think about a perception before understanding it). However, something as objectively simple as a hand picking up an object cannot be understood immediately and unequivocally. The act of picking something up is undeniably objective, but it has no meaning, unless the context is known (*why* the object is picked up). In any event, the act must be perceived by an observer who may read many things into the act of picking up the object. If the object is a hammer and the human is known to be a carpenter, that is one thing; if the act of picking up the hammer is associated with the human's expression of rage, that is another thing.

This is the factor of *meaning*, in that objective phenomena cannot be completely understood unless someone is around to interpret the phenomena. Consequently, completely objective phenomena do not exist, because objective phenomena cannot be divorced from the human. Of course, objective phenomena could be those that can be recorded only by instrumentation, like magnetic radiation, but that would limit the definition of what is objective to only a few things, and in any case radiation, like other "objective" phenomena, must be interpreted by a human, which puts things back to "square one." Any definition of objective phenomena cannot exclude the human, which is fortunate because otherwise there would be little role for the human in measurement.

THE SINGULARITY OF SUBJECTIVISM

Why then consider subjective methods as a category distinct from objective ones? It is because subjective methods have peculiarities of their own.

In subjective data collection, the subjects' consciousness of what they report may serve as a constraint on what is reported. Subjects may not be able to describe how/what they feel. They may report only partially and may select what they decide to report, thus "slanting" what is reported.

On the other hand, subjective data added to objective data may explicate the latter. It was common practice in World War II bomb disposal (and it may still be so) for the operator who dismantled the bomb to report the activities he was performing to a remote team member, so that, if the dismantling was fatally unsuccessful, the source of the difficulty could be analyzed later.

There is a tendency to believe that objective data, or rather the meaning of objective performance, is clear, because the data are objective; that is not so. The meaning of even objectively recorded performance may not be obvious. (See the preceding "hammer" example.)

Meaning in objective human performance is more clear-cut, but also somewhat partial, because only half of what goes on during performance (the physical part) is reported. Does it distort the objective data, because nothing internal is correlated with that performance? Objective performance presents only what happened; if the task scenario is complex or confused, the subjective report may suggest context, and thereby clarify the objective data.

If objective performance is not accompanied by interrogation of the subject, that objective performance presents only an incomplete, hence distorted view of what has happened. The behaviorist will say: Why does it matter that behavioral data are not gathered at the same time as objective data? What counts is the objective performance, and not what the subject thinks about that performance.

This may be true in highly proceduralized tasks, but in more cognitive, complex, less well-defined tasks, what subjects think about what they are doing may well determine (at least partially) the overt performance. Do the behavioral concomitants of the physical performance make any real difference to what the subject actually does? It seems so, but how much is a matter for research. The fact is that objective and subjective elements are so intermingled in human performance that a definition of one that does not take account of the other is deficient.

One of the major purposes of HF research is to determine how the human reacts to technology. This issue is studied objectively on a molecular level (e.g., response to icons, menus, type fonts). Researchers do not study how internal status affects people's reaction to technology as a whole. Part of that reaction is subjective. There is, for example, some anecdotal evidence that prolonged utilization of computer technology, as in the Internet, creates a behavioral (possibly even a physical) dependence in the user, which may be expressed in changes in information processing and decision making.

WHAT ARE THE SUBJECTIVE METHODS?

The subjective methods to be discussed in the remainder of this chapter have been categorized (somewhat arbitrarily) as *observation*, *self-report*, and *cognitive*.

In observation, the human acts as the instrument to describe the performance of others and of objects, as in inspection.

In self-report, humans record their own internal status. Self-report includes any of the following outputs: physiological conditions, as in a patient reporting symptoms to a physician; attitudes and opinions about objects and phenomena, as in a survey such as is described in chapter 3; judgments about the nature of internal and external phenomena; and decisions based on those judgments.

Cognition involves detection, identification, classification, comparison, and scaling of objects and phenomena, usually leading to some sort of evaluative decision. These internal functions also support observation and self-report.

These methods and functions overlap, of course; when people report internal status, such as stress level, they must also observe that status, which requires detection, identification, and so on.

In the remainder of the chapter, the discussion is organized around the same categories as in chapter 5.

OBSERVATION

Description

The reader may ask why attention is given to observation, which is not often utilized as a method in the general HF literature. On the other hand, observation of operators performing predecessor system tasks is a part of most system-related studies. Observation is the most "natural" of the measurement methods, because it taps a human function that is instinctive. If an individual is conscious and sighted, that person observes, even though, as in the Sherlock Holmes stories, inadequately. A layperson's observation is directed by that individual's capability, situation, and interests, but it is not ordinarily directed in the same way as is observation as measurement. Observation in measurement is goal directed: to understand a performance whose meaning is not immediately obvious. Thus, there are at least two types of observation: the more common casual observation, which is undirected by a specific measurement goal, and directed observation, which has a definite measurement goal.

The goal tells the observer what to look for or at. Except in the case of a demonstration that is being observed, the observer has no control over what is manifested by the objects or people being observed. Those being observed are often unaware that they are being observed, although, again, there are exceptions. Where this awareness exists in those being observed, self-consciousness may make their performances somewhat artificial and mannered.

The report of what is observed is made by the observer, which makes the observation subjective, even though it may appear objective because the observer is not personally linked to the subjects of the observation or to what is observed. This aspect, that the observation is made by and transmitted through a human, is responsible for the suspicion felt by many researchers for the method. It is well known that there are significant individual differences in observation ability. Most researchers have not been trained as observers, so that unless the performances they are observing are quite overt, with a known scenario, such as a formal procedure (i.e., known to the observer), there is likely to be an unknown amount of error in the observation report. It is proverbial, for example, that most witnesses are poor observers.

Scenarios describing what is to be observed vary in the amount of their detail. The minimum scenario is the knowledge, usually available prior to an observation, of the general context of what the personnel being observed are doing (e.g., what is being observed is a cocktail party, or a team of nuclear power control operators). The *details* of what the observed are doing are often unknown, and therefore the details are what the observation is designed to discover. The desired information (the observational outputs) is described as categories on a check sheet (or its electronic equivalent; e.g., number of interpersonal contacts, number of telephone messages, number of times questions are asked). Some of this information can be gathered automatically (e.g., number of telephone messages), but other items of information might be difficult to instrument, and there is also the matter of relative cost, time, and difficulty of automatic measurement. This may cause a researcher to use observation rather than instrumentation.

In the complete absence of a scenario, observation is likely to be unorganized. If nothing is known about what is to be observed, then categories of data to be collected (these are actually hypotheses about what is important in the activity to be observed) cannot be established in advance. That is one reason why self-report statements of internal states (e.g., the response to the question, "How do you feel?," of "O.K., not so bad") are likely to be somewhat vague.

Meaning is important to all measurement methods. Every measurement situation has some meaning, relative to what personnel are doing or are expected to do, embedded in that situation. That meaning derives from the purpose of the individuals observed and the purpose for which the measurement is being made.

Purpose

The purpose of observation as a methodology is to secure information about what has happened and is happening—the status of personnel and the mechanisms with which they perform and/or the characteristics of objects. When objects are involved, the method is called inspection.

Because of the constraints under which observation of others (i.e., noninterference) ordinarily occurs, the purpose of the observation is limited to determining *what* those who are observed are doing, how they are doing it, and their status at the time of observation. Because during the observation it is not permitted to manipulate the observed, or influence their behavior, there is little opportunity to discover *why,* unless an individual has access to the details of the procedure followed by the group. Observation of human subjects is essentially a "hands off" process; that is why information about context is so important. It is possible to interview the personnel being observed later to secure information about the how and why, but the methodology then involves something more than observation.

Use Situations

There are no limitations (other than those of subject privacy and physical access to subjects) as to where and when observations can be employed. However, observation is primarily employed in the following HF situations (there are, of course, other non-HF, psychological research situations in which the method is used):

1. Demonstrations of task performance. In the early stages of new system development, personnel performing tasks on a predecessor system may be observed to assist in the development of a task analysis for the new system. If jobs are to be redesigned, then observations may be made of how personnel perform their original jobs in order to ensure that the new job design incorporates the old functions.
2. Observation may be of use during testing of a preliminary design of a new system. Developmental questions, such as whether a particular design will permit certain tasks to be performed efficiently, may be studied by putting subjects into or in contact with a prototype of the new design and observing their performance. For example, studying how well manual maintenance tasks are performed in extremely restricted areas such as submarine spaces. The prototype test that exposes preliminary designs to potential consumers for their approval may involve observation of their responses as part of the prototyping test. In the entertainment industry, observation is an essential function in the development of new plays, musicals, sport, and so forth. Observation may be used in the training or evaluation of student performance.
3. Observation of ongoing operations during the first operational trials of a new weapon system, such as a destroyer.
4. To elicit information for development of an expert system or for tutorial purposes. For example, a physician will demonstrate to an observer a diagnostic process to be incorporated in an expert system, or will demonstrate to interns during patient "rounds."
5. To determine the status of objects as in the inspection of objects.

The previous examples emphasize system development situations, because observation is rarely used as a method in conceptual research, which makes primary use of experimental methods. However, in ordinary life, observation is a basic method of securing information and making decisions.

A job performed according to step-by-step procedures reduces the need for its measurement by observation, because the procedure defines the "what" of the job. The more unstructured the job and the more personnel performance depends on contingencies, the greater the utility of observation. In such jobs, the "what" of events may be unclear and must be determined. An example of an unstructured job is the air traffic controller's work, which often requires nonroutine actions based on the application of general principles to emergency conditions that cannot be proceduralized, because they cannot be completely anticipated.

It goes almost without saying that the actions and events being recorded through observation must be overt; if covert behaviors are involved in what is viewed, they can only be inferred, unless the observation is followed by interviews with the personnel observed.

Questions Asked

These questions are of two types: (a) What questions is the study intended to answer? (b) What questions can the observation methodology answer? Comparisons of these two questions determines whether observation is selected as the measurement method.

Observation can indicate what, when, where, and how actions are being performed. These are descriptive answers only. Observation cannot determine which performance is better (although an expert observer can make this judgment); whether the system performs as required; what the variables underlying performance are. These are not amenable to observation (although observation can aid in making these judgments), because it cannot manipulate what is observed and cannot question personnel directly at the time of performance. All that an individual can get from observation is the answer to the general question: What is going on? Whether this question should even be asked depends on how much prior information is available about the tasks in question.

Simple description is only a starting point in measurement. Individuals must know what goes on before they can ask why it is going on, what the factors are that affect what is going on, and what the mechanisms involved in the performance are. Levels of knowledge are involved: *what* is simpler to discover than *why* and *how*.

Selection of observation as the method to be employed depends on the answers to the following:

1. What questions does the researcher wish the observation to answer?

2. How much is known about the activity being observed? The more that is known in advance, the more veridical the observation will be. On the other hand, if a great deal is known about the task, the amount of additional information provided by the observation may not be worth the effort. Proceduralized tasks are not good candidates for observation, because the procedure defines the task very completely, leaving little room to look for variations. However, because of proceduralization, observation can be used to record errors (deviations from that procedure). Contingent tasks are better candidates for observation, because only observation will tell researchers what the most frequent contingencies are, and the most likely subject responses.

3. How overt/covert are the behaviors to be observed? This may be the deciding factor in the use of observation, because if behaviors are highly covert, there may be nothing to observe.

4. Can the observation stand alone, or must it be associated with another method, such as the interview? What alternative methods (e.g., instrumentation) can be used in place of observation? Then researchers must ask about the relative accuracy of observation and such alternatives, and the relative cost of observation (time, money, resources) compared with the alternative methods. The selection of observation as a measurement method may involve a series of trade-offs.

5. How many observers will be needed? How much training do they need to perform the observation? What level of agreement among observers will the researcher accept as indicating veridicality of the observation?

6. What is to be recorded by the observer? Is there a recording form with performance categories to be observed? Sometimes the behaviors that can be observed will be more diverse than the observer can record at one time (unless audio-video recording is also used, but in that case who needs a human observer except later, when the data are being analyzed?). Are the observational categories self-explanatory or must procedures for utilizing them be explained to the observer? Do the observers need to be pretested to eliminate potential problems?

7. What kinds of data analysis will be performed? How are the data to be utilized? (This last question should be asked very early in the process of selecting any method.)

Assumptions

The use of observation assumes that what is observed represents the essential nature of what is happening. (This is not the case in experimentation, which is artificial to the extent that it "arranges" what is to be measured.) As was pointed out previously, observation makes the observer equivalent to measurement instrumentation. Use of observation therefore assumes that observers record veridically; if they do not, the entire method is invalidated. The validity of observation can be tested in part by employing several observers and determining the reliability of their responses.

Procedure

Some of the procedures to be followed in using observation have been anticipated. It is necessary to decide on the questions to be answered, and whether observation will provide answers to these questions. (This process is required of all methods.)

The observation is focused by the questions to be answered; these are converted into observational categories that represent the researcher's hypotheses about what is important in the performance to be observed. For example, if the researcher's hypothesis is that messages are unequally distributed among team members, then the observer will have to check the frequency of messages for each team member.

Researchers cannot manipulate the personnel or object observed, but they can focus the observers by pointing out the things to look for. The conditions of observation must be carefully assessed and followed. Because humans are the instruments for performance recording, they must be calibrated (as all instruments are calibrated) by practice and testing to determine the reliability of their responses. This reliability does not ensure the validity of the observational process, but because what is observed is assumed to be the essence of the performance, validity can be assumed if sufficient observer reliability can be achieved (see Boldovici et al., 2002, Appendix C). The level of reliability required—98%, 90%, 85%, and so on—is the decision of the investigator, but squaring the achieved reliability will indicate how much observational variance is accounted for.

Advantages

Observation has a few advantages. The most important of these is that whatever and when actions are being performed, someone is observing these actions: the performers themselves and others involved in the performance whose observations and recollections can be tapped.

Because the questions it will answer are descriptive only, observation in research is useful primarily as a predecessor and an adjunct to more sophisticated methods like the experiment. Observation is rarely used in conceptual (i.e., nonsystem related) research, because the researchers usually create their own measurement situation. However, if the measurement task is based on an operational one, there may be an advantage to observe the operational task first.

Observation has an advantage in the sense that the act of observation does not interfere with or influence the performance being observed (unless those being observed are aware of the observation).

Disadvantages

However, the inability to manipulate what is being observed, or to ask questions of the observed during the measurement, is a major weakness, because it limits the questions that can be answered with the method.

Outputs

Observation can provide a description and characteristics of what occurs, including the *frequency* of and the *sequence* in which performance occurs. If errors can be observed, then these can be described and counted by the observer. Performance time can also be measured using a timer.

Summary

Observation is more often used in a supporting role than in a primary role.

INSPECTION

Description

This is a method commonly employed in factories and in maintenance facilities to ensure quality control and performance adequacy of components and equipments. The individual (inspection is almost always an individual task) peers intensely at a small component (e.g., a circuit board) and compares it with a photograph of the component as it should look after assembly. The component can also be activated and its response measured. Inspection is also applied in using design guidelines in evaluation of the human–machine interface. Comparison is an essential element of inspection, as is the existence of a standard that may be concrete and detailed or vague and general. The end purpose of inspection is acceptance or rejection of the component, if discrepancies are found between it and the standard.

If there are degrees of observation, inspection is the most intensely focused form of observation, so much so that it can be considered a new method. Observation in its original, natural, and primitive form is simply perception and does not involve a formal comparison of what is observed with a standard, although all observation may depend on an expectation of what is to be seen. As a result, an anomalous element in what is observed can cause the observer to note a discrepancy from that expectation. The expectation serves as a crude standard.

Observation, attention, and inspection are interrelated. Observation, which relies on perception, becomes inspection when it is concentrated on an object or phenomenon. The mechanism mediating inspection is the human's attention applied to perception. The result can be likened to a narrow beam of light that illuminates the object of the inspection. Inspection is almost always molecular, whatever its venue, because of the narrowing focus of attention.

Another difference between inspection and observation is that in inspection the object being observed can be manipulated in position, in lighting, or in magnification (as in a microscope), whereas in general observation this is not possible.

Inspection need not involve electronics only; any object can be inspected. The object of the inspection can be a painting, a mathematical equation, or anything about whose characteristics there is some question. The essence of the inspection is a comparison of one object with another or of that object with a physical or mental standard. The standard is often or even usually visual, but a musician in tuning a musical instrument may perform the same function aurally. The standard against which the object is being measured may be physical, like a photograph, or a comparison object, or it may be mental, as when a physician examines a patient, compares the patient's symptoms against a mental catalogue of possible illnesses, and makes a differential diagnosis. Inspection is always close examination, directed by a clear goal; general observation is not goal directed; when it is, it becomes inspection. Spectators at entertainment functions act as quasi-inspectors. Inspection is therefore a common human activity; inspection in research or production (the HF interest) is only a subspecies of the more common activity.

The process of evaluation requires the definition of a standard, which until now has not been characteristic of the other measurement methods discussed. True, general observation involves a standard in terms of an expectation derived from experience, but such an expectation is rarely focused or formal, and often is not goal directed. The existence of a standard plus the evaluation of the object being inspected now makes decision making an integral part of the inspection process.

One of the most difficult aspects of the inspection process is the development of a standard, because this is a complex process requiring determination of the characteristics of the object inspected that determine the performance of that object. For many repetitive tasks, like quality control factory inspections, the standard may have already been developed, but in other measurement situations the standard does not yet exist and must be created, which makes for a certain amount of trial and error.

Standards are not to be found in experiments and in other conceptual research directed at the confirmation of hypotheses related to variables. Standards are an integral part of system development, because the question—does the system perform as expected or desired?—is an integral part of system development. The standard may be very precise, as in inspection of components, or it may be rather gross; it is associated largely with physical entities, system development, and operation, but may also exist in personnel as a half-formed mental impression in examining internal conditions. A standard always implies an intent to evaluate something.

Inspection has been categorized as a subjective method, because, along with observation from which it is derived, the primary measurement instrument is the human, the inspector (although the inspector may use an instrument as an aid). At the same time, inspection differs from ordinary observation, in part because the point of the inspection is to perform an evaluation. Ordinary observation, even for measurement purposes, usually does not involve highly detailed inspection, and certainly not evaluation.

To the extent that observation in work and nonwork activities involves some inspection elements, inspection is ubiquitous, to be found not only in connection with equipment or artificial products, but also as an ordinary part of life activities. For example, soldiers are often inspected by superiors to ensure their readiness for duty. When people meet someone for the first time, they inspect the person, and the first output of the inspection is an immediate decision: male or female. Inspection involving such fundamental characteristics is almost always unconscious, which differentiates it from inspection as deliberate measurement. In all such inspections a standard exists, although it may have been learned in the remote past.

One may also find inspection as a process involved in measurement. For example, after an experiment, the researcher inspects the data before performing a statistical analysis.

Purpose

The purpose of inspection is to evaluate an object, an event, a person, to ensure that it or the person satisfies some standard. The evaluation requires a decision about this comparison.

The inspection involves the collection and processing of information, but this process is not the purpose of the inspection; the evaluation decision is.

All measurement involves the collection and processing of information, but the specific purpose of any particular method is to arrive at a decision based on that information.

Use Situations

In its common industrial form, inspection is part of the production phase of system development or maintenance phase of system operations. However, inspection is performed in many more venues. For example, the HF design specialist in evaluating a human–machine interface (HMI) or human–computer interface (HCI) is performing an inspection, the standard of which is called a design guideline, which lists the relevant characteristics of the HMI/HCI and their required appearance or performance. This kind of HF inspection is more arduous than industrial inspection, because the criteria in the former are likely to be less precise (see Meister & Enderwick, 2001). Inspection is also performed in hospitals (on patients, of test data), in museums (of artifacts), as part of artistic and musical performances, in sports entertainment, and so on.

Questions Asked

Several questions must be asked before inspection is completed:

1. What are the characteristics of the object being inspected that determine its effective performance? The object may have many characteristics, some of which are not very relevant to performance (e.g., the color of a workstation).
2. Do the relevant characteristics of the object under inspection accord with those of the standard?
3. If discrepancies are found between the object under inspection and the standard, do these invalidate effective performance? If what is inspected is an object, if the discrepancy is great enough, then it will be rejected. If what is inspected is a human (e.g., a patient), a course of remediation will be tried. If what is inspected is a human performance, then a discrepancy will usually require additional practice.

Underlying Assumptions

The single most important assumption underlying inspection is that the object under inspection may not be as expected or desired and the available standard depicts correctly what the object should be.

Procedures Employed

The following steps are performed:

1. Determine which object is to be inspected (in most cases this is determined before the inspection is conducted).
2. Determine which aspects/characteristics of the object are relevant to the question: Is this object adequate? (Also predetermined, in the form of a physical standard or a mental concept available generally, or only to certain specialized individuals, e.g., the physician).
3. Access the standard (physically or by recall).
4. Select from the standard the relevant characteristics of the object to be inspected.
5. Compare the object's relevant characteristics with the appropriate characteristics of the standard (certain tests, physical, mechanical, medical, may be required to perform the comparison).
6. Make a judgment between the object and the standard: same or different. If different, ask the question: In what way, and are these differences significant?
7. Accept or reject the object being inspected.
8. If a fix of the offending characteristics is possible, then indicate the nature of the fix.

Advantages

The term *advantages* may not be appropriate for such an innate, required, human function. The methodology exists and therefore in specified circumstances (e.g., production quality control, patient diagnosis) must be utilized. In this respect, inspection differs somewhat from other behavioral measurement methods that need not be utilized, if a more effective method can be found.

Disadvantages

The inspector's knowledge of the relevant characteristics of the object being inspected may be less than optimal (e.g., defective and incomplete). The standard of comparison may be vague or lacking the relevant characteristics. The inspector may make the wrong accept–reject judgment.

The most critical part of the inspection process is the development of the standard, which is often a lengthy, complex, trial-and-error process.

Outputs

There is no direct, immediate physical output of the method, only a decision. If accepted, the object, the performance, the person remains, and functions; if rejected, it is discarded, or, if a human, is remediated by medical methods; if it is a defective human performance, it is remediated by training. The output of the inspection process is often the preliminary to further activities, as in medical diagnosis or equipment maintenance.

Summary

In certain circumstances (e.g., industrial quality control or medical diagnosis), inspection is absolutely required. The process is subject to error; hence the mechanisms involved in inspection (the standard of performance, the inspector's level of expertise) must be understood, and the inspectors themselves inspected to ensure their adequacy. Inspection is usually part of and supports a superordinate process like medical practice, industrial production, entertainment development, or archeological discovery.

THE INTERVIEW/QUESTIONNAIRE

Description

With the interview and questionnaire a change occurs in measurement methodology. In observation and inspection the object of investigation could not be queried during the process and so could not provide information about the object's internal status or what the object, if human, had perceived of external events. With the interview and questionnaire, those personnel who are the objects of observation and inspection can now be interrogated and can respond. They can even be manipulated, which leads to the experiment.

The interview and questionnaire are lumped together because they have essential elements in common. In both, the human is asked questions and invited to respond. In the interview, the questions are received aurally; in the questionnaire, they are asked in written or electronic form. The interview is usually (although by no means always) one (interviewer)-on-one (interviewee); the questionnaire is almost always administered to groups of respondents. The interview may be formal, with questions in a predetermined sequence, or free ranging with the choice of topic contingent on what the respondent said previously. The questionnaire is (must be) always formal, because it is presented to groups, each member of which must receive the same questions, phrased in the same way; otherwise, their responses cannot be combined. The questionnaire is often used as a substitute for the interview, because the interview requires much more time and is more labor intensive than the questionnaire. Both interview and questionnaire can be administered in many venues, over the phone, and so on.

A number of assumptions underlie the interview/questionnaire. The first is that humans can validly describe their reactions to stimuli. This assumption is common to all subjective methods, but the assumption is somewhat different in the interview/questionnaire. In observation and inspection, the observer records the performance of *others*, either some human performance or the performance characteristics of an object. The validity of the observer's perception can be cross-checked by reference to the perception of others. In the case of the interview/questionnaire, the humans/observers emit their own responses. These responses can be, as in observation/inspection, the perceptions of what others do, or they may describe the respondents' own behavior or feelings. Cross-checking of the interview/questionnaire's responses is possible where the observation is of something overt; where the observation is of the respondents' own condition, it cannot be cross-checked, except in medical practice.

Observation of internal status consists of two phases: observation/recognition of that status, and reporting of that observation. The two phases are independent, although related. It is at least theoretically possible to recognize internal status (How do I feel today?), but lack the words to report that status. What this means is that although veridicality of the internal status report is assumed, there are degrees of veridicality. The pragmatic assumption is made that interview/questionnaire responses dealing with that internal status are reasonably veridical. Otherwise it would be impossible to make use of the data, because there is no external standard, as in inspection, to determine correctness of the subject's response. Internal reality is always only an approximation, with which researchers have to be satisfied. (There are other realities: interpersonal reality, or what people do, and technological reality, or how machines function. These realities can be accepted without question because their stimuli can be sensed by others; but internal reality is always "dicey.")

Another assumption is that the humans' internal status can influence their own and, through this, system performance. This assumption is the rationale on which the interviewer/questionnaire developer asks for data about an internal condition.

The fact that reality can only be approximated has significant implications for measurement. Measurement is an attempt to depict that reality. If the notion is accepted (as it must be) that approximation is the best that can be done with measurement of internal status, then the question arises: How much approximation is "reasonable"? Unless conventions of reasonability are developed, any set of responses may, or may not, be considered reasonable.

One of the difficulties of using "reality" as a referent for measurement is that reality has many dimensions. What constitutes reality depends on *whose* reality is being talked about; the aspect (e.g., machine operation, combat, inner feelings) the measurement is supposed to refer to; the level of detail with which that reality is contemplated, and so on. All of this is philosophical, of course, but it has practical significance in terms of determining whether this is the "truth" of what has been measured. Almost all measurements, other than those conducted in the operational environment, are surrogates for a particular reality. On the other hand, it is questionable whether most HF researchers even think in these terms.

The researcher can establish the bounds of what is reasonable. In the case of the experiment, the significance of difference level of .05 probability is generally considered an acceptable limit in "proving" that the data results did not occur by chance. It is not clear that any similar bound of reasonability can be established for internal status reports. Statisticians have probably developed quantitative methods of ascertaining the validity of internal subject reports; but researchers do not have to believe in these.

Of course, why impose any bound on a reasonable approximation of veridicality? It could be maintained that any performance value, any set of responses provides some approximation, some picture snapshot, of veridicality, assuming that there has been no egregious bias in the measurement process. But for the sake of order in the measurement, somewhat arbitrary limits must be established. The fact that any measurement output is only an approximation emphasizes the importance of the researcher's logical deduction from that measurement output. If, to use a hypothetical example, 80% of the truth is obtained from any measurement process, then logic, experience, intuition, and so on, must of necessity supply the remaining 20%.

If veridicality is only partial and must be supplemented by the researcher's expertise, confidence in reality is often shaken by an unacceptable ambiguity.

The solution is to assume that all data satisfying previously specified quantitative minima are *ipso facto* veridical and reveal reality. That assumption is bolstered by repeating the same study and performing related studies; researchers look for measurement outputs in the new studies that approximate the previous ones. This is the preferred (although not necessarily usual) way of assuring validity. The picture of reality provided by measurement is built up slowly and incrementally (with some unknown amount of error).

In all this process it is necessary to rely on the human interpreter of data, which means that the subjective contribution to measurement becomes increasingly important.

The mechanism of interrogation in both the interview and the questionnaire is the "word," which presents many problems: Do the words (questions) incorporated in the interview and questionnaire accurately represent what the interrogators intended to ask? Does the interviewee/questionnaire respondent accurately understand what the words mean? In reporting internal and external status (how humans feel, what they understand of their internal and external condition) can humans efficiently recognize that status (i.e., the stimuli describing that status) and communicate their recognition of this status accurately?

It is apparent that there are many opportunities for error. Words have connotations that modify their plain sense. Respondents, being usually a product of the same verbal culture as the interview/questionnaire developer, also respond to these connotations. These, in turn, modify their verbal responses, which are also in words with their connotations.

The subject can respond in terms of the present, past, or future (anticipated actions). In the past, there is the possibility of memory loss. In the future, there is the possibility of a change in the respondent's intentions. Responses to questions can be influenced in several ways: (a) The questions themselves are either too simple (do not contain sufficient relevant parameters) or too complex (contain too many parameters for an unequivocal answer) to elicit responses that represent what the subject had in mind. (b) It is an interactive interview in which the questions asked by the interviewer are contingent on the interviewee's previous answers. Or, (c) the interviewees can question the interviewer, if they do not fully understand a question.

Questions may ask for qualitative (verbal) responses only, or quantitative answers. The manner in which a question is asked can determine the form in which the respondent answers (e.g., multiple choice answers or more detailed, written replies). Answers may also be skewed by an unrecognized bias included in the question (e.g., When did you stop beating your wife?). Even the sequence in which questions are asked may produce an unanticipated effect in the respondent. The interview may follow a performance, in which case the interview is called *debriefing*. In police circles, the same would be called an *interrogation*. When suspects are interviewed, they may make a voluntary statement that is a confession.

The interview may be conducted as part of a demonstration of skill. It may be formal (as most interviews are), directed by specific questions developed previously; or the interview may be quite informal, involving questions developed during the interview, as the topics of the interview change direction.

It is apparent that the interview as a form of self-report assumes different guises, occurs in various settings, and provides different types of information. The interview may be directed to secure information about a specific individual (e.g., as in medical diagnosis) or the individual as a member of a group (responses as a function of gender or socioeconomic class). The interview may be of an individual or in a group (i.e., a focus group). The information content researchers attempt to elicit may be many things (e.g., preferences for deodorants, salary situations, conceptual structure), all of which is lumped under the term *information.*

Purpose

The individual or group being interviewed is presumed to have some information only that person or group possesses. The purpose of the interview/questionnaire is to elicit that information.

Use Situations

The interview may be conducted at any time that some information, which the respondent(s) has, is needed, before, during, and after a test, during system development and in system operations. The decision concerning whether or not to conduct an interview or questionnaire at any particular time depends on circumstances, such as the immediacy of the need for the information or circumstances that might bias responses.

The interviewer or questionnaire developer is an integral part of the measurement process. In the interview, the respondent reacts not only to the words asked, but also to the interviewers' expression and their "body language." In administering the questionnaire, researchers may not be present, but their choice of words is extremely important, because that is all the subject can respond to. The development of the interview/questionnaire (as reflected in the choice of words) is as much a creative process for the researcher as the poet's choice of words for a poem—and equally mysterious. There are detailed rules for writing questions (see Meister, 1985, or Babbitt & Nystrom, 1989a, 1989b), but even so, the process is highly creative. That process requires the interview developer to *anticipate* subject responses to the measurement instrument.

Questions Asked

The interviewer/interrogator must indicate specifically and precisely what the general topic is. The following are examples of the general types of questions that may be asked:

1. What does the respondent know about a particular skill?
2. How does the respondent perform a task?
3. Why did respondents do what they did at such and such a phase in system operations or in a particular test situation?

4. What did the respondent see or hear of someone else's behavior or a phenomenon?
5. What is the respondent's attitude, opinion, or feeling about some piece of equipment, system aspect, concepts, or practices?
6. Given such and such a hypothetical situation, how would the respondent react?
7. What is the group consensus about a specific problem or issue?
8. How do the respondents feel about themselves (as in a psychiatric interview), about their physical/emotional condition?

Underlying Assumptions

Certain assumptions have already been discussed. These may change somewhat as a function of the type of information being collected and the nature of the subject sample. Where respondents are experts, like highly skilled physicians, researchers might give greater credence to the information they provide than they would to, say, a college sophomore. This would, however, also depend on the nature of the information being collected. If attitudes toward "rock" stars were being collected, then college sophomores might be given much more credence than a physician sample. Obviously, there is some doubt even with recognized experts; some experts are more expert than others.

Procedures Employed

If an interview is formal, then it is presumed that the questions asked will have been developed beforehand and even pretested. This is even more true of the questionnaire. The questionnaire is always formal; the interview may be more or less formal. From the latter, sometimes only major themes or questions will be developed and the strategy of conducting the interview is to begin with these questions and then ask further, more specific questions, depending on how the interviewee responds.

It is undesirable to include in a single interview or questionnaire item more than one theme, because several themes in one item can confuse the respondent and make it difficult to determine to which theme the subject is responding. There are times, however, when the subject matter is inherently complex, consisting of multiple parameters. The survey in chapter 3 is an illustration.

In real life, concepts do interact and those interactions must be considered in asking questions. If the single theme paradigm is followed, then interviews/questionnaires dealing with abstract material are likely to be artificially simplistic, as will be the responses to them. One way of circumventing the single theme provision is to introduce each of the interactive concepts first in individual items, and then to include in further items the themes in interaction. Then it is possible to compare responses to the separate and combined themes and determine the effect of the interaction.

The problem with this solution is that it extends the interview/questionnaire significantly, often beyond the willingness of potential subjects to participate. Consequently, in a very long measurement instrument, subjects may tend to respond superficially.

There is no easy solution to the problem. Often, as was the case with the chapter 3 survey, the researcher has a choice among undesirable alternatives: to cover only a few limited themes in the instrument (when, in fact, many themes are important), and thus make respondents more willing to participate; or cover the full range of themes and risk reducing the number of participants to a less than desirable number. Because of the assumed capability of the chapter 3 respondents, the single theme provision was ignored. This particular problem does not arise with simple physical motor or perceptual materials. It is a problem only with complex conceptual material.

Almost certainly the first effort at wording a questionnaire requires revision, and it is particularly necessary for the questionnaire developer to anticipate possible misunderstandings. It is usually good practice to repeat a question or theme at least one time, varying its wording; this gives one a reliability check on what the respondent says. A pretest of a questionnaire is highly desirable, because it inevitably reveals items with a potential for misunderstanding.

Researchers must decide on the subjects of the interview/questionnaire. This depends on the theme of the instrument. If the purpose is to elicit specialized information, then the respondents should be subject matter experts. If respondents are selected because of their membership in a cohort (e.g., aged, disabled, etc.), then selection depends on their classification. If a researcher is interested only in feelings, attitudes, or opinions of people in general (e.g., potential voters), a sample representative of ethnic, education, and socioeconomic diversity must be selected.

Researchers should consider how the data will be analyzed before administering the interview/questionnaire; this fact may change the way questions are worded. This is particularly true if the questionnaire contains any quantitative scales.

The kind of responses the subject will be asked to make to the test instrument is important. In questionnaires the respondents cannot be expected to write out lengthy answers (which would, of course, be most desirable), because they will simply not respond this way.

On the other hand, very simple responses (yes–no; frequency, etc.) will probably supply overly simple answers that are often less than adequate for complex material. The interview/questionnaire developer needs to consider this complexity parameter when selecting a response format, or in deciding whether to use an interview or a questionnaire. This, again, is one of those judgmental issues for which there are no standard answers.

The most important practical question for researchers is when they would prefer to administer a questionnaire rather than an interview. The questionnaire enables the researchers to collect a great many more responses in a much shorter time than the interview will permit. For complex topics, direct one-on-one interviews are to be preferred to formal, mailed questionnaires, because in the former researchers can step off the beaten topic, explore interactive aspects, and provide explanations if the subject becomes confused. Unfortunately, one-on-one interviews are tremendously costly in money and time, because subjects must be queried in sequence rather than in a single broadcast mailing. Another difficulty is that qualitative interview responses are more difficult to assemble and quantify than quantitative questionnaire replies. However, such combinations can be performed.

Advantages

The greatest advantage that self-report methods have is that covert phenomena can only be investigated by asking the subjects to reveal knowledge only they possess.

Disadvantages

The disadvantages of the interview/questionnaire are those of self-report techniques generally. Any subjective response may depend on the respondent's memory, which may falter; the subjects' comprehension of what is requested of them, which may be only partial; the connotations of words, which may produce ill-considered responses; the subject's difficulty in accessing internal concepts and conditions. The subject may have difficulty expressing knowledge or internal impressions, when the interview is conducted informally and formal response categories are not specified. On the other hand, any response format controls and therefore skews, to some extent, what the respondent wishes to say.

Outputs

An informal interview may produce statements that may require content analysis. If the subjects are also asked to scale their responses or to perform certain measurement tasks as part of the self-report (e.g., paired comparisons, ranking, rating), then a quantitative value becomes available, so at least frequency statistics can be applied. If there is more than one type of respondent (e.g., experienced vs. novice) or two different situations are investigated, these can be compared by applying significance of difference statistics.

Summary

Investigators are never completely satisfied with self-report techniques, because so many factors may impair their usefulness. On the other hand, no measurement specialist should ever be satisfied with the measurement results achieved with any method, because any method is always dealing with only a partial representation of reality. Nonetheless, in many situations, particularly involving covert material, there is no alternative to the self-report.

JUDGMENTAL METHODS

Description

The term *judgmental* is used because the end result of using these methods is a conscious decision on the part of the respondent to discriminate between two objects or conditions, to select one response or the other, and so on. The psychological processes involved in these methods are essentially the same as those found in all perceptual and cognitive behaviors (i.e., detection, analysis, recognition, classification, comparison, scaling, and selection). These become judgmental methods when they are formalized and developed overtly (externalized) to express the subject's internal condition.

Among judgmental methods are the psychophysical methods, which reflect the subject's perception of external qualities, such as the threshold of discrimination of size, weight, or distance. Little will be said about psychophysical methods and about signal detection theory (Green & Swets, 1966), which describes the ability to discriminate the occurrence or nonoccurrence of physical signals. Those who are interested in the extensive literature on psychophysical methods and signal detection theory should refer to R. W. Proctor and Van Zandt (1994) and to R. W. Proctor and J. D. Proctor (1997).

There are two ways in which humans can express their reactions to technology: first, by objective performance in relation to equipment and systems (e.g., how icons and menus affect the efficiency of the operator's performance); second, through the subject's attitudes toward and concepts of that technology. For example, do subjects approve all aspects of technology; are there any technology aspects that are undesirable; how do humans define the impact of technology on them personally?

Some readers may feel that only objective performance relative to technology is important, but this is insufficient. If technology does not increase the individual's perception that life is better with a certain aspect of technology than without, then technology should, if it can, be reined in; and those aspects of technology found offensive should be eliminated, if they can be. HF has a responsibility to determine human attitudes toward technology. Judgmental methods are needed to determine these attitudes.

Judgments vary on a continuum of complexity. The psychophysical methods deal with single dimension stimuli. The judgments in internal self-reports often involve stimuli with multiple dimensions that vary in complexity, concreteness, and manner of expression (e.g., qualitative or quantitative). The judgmental self-report methods apply numeric values to self-reports, sometimes in interviews, but more often in questionnaires.

Subjective phenomena (internal status conditions) are thus given the appearance of externality, which enables the researcher to cross boundary lines between the behavioral and physical domains.

The methods permit the subject to make the following judgments:

1. *Frequency*, for example, how many times in a period does the subject perform some action or does some phenomenon occur?

2. *Sequence*, or the arrangement of conditions (e.g., stimuli, actions) in order of their occurrence.

3. *Classification*, for example, the judgment that Condition X has certain characteristics or is related to or part of Condition Y. For example, at a physical level, stomach pains may be associated with diarrhea; at a conceptual level, relations can be indicated by sorting cards listing certain factors involved in making a decision, or equipment failure reports can be sorted on the basis of the failure cause. The highest form of classification is the development of a taxonomy of human functions, such as those by Berliner, Angell, and Shearer (1964) and Fleishman and Quaintance (1984).

4. *Comparison* of conditions, objects, stimuli, and so forth in terms of some attribute like size, weight, frequency, or importance.

5. *Choice* enables the respondent to select among alternatives (e.g., a multiple choice test). Although all the judgmental methods require implicit judgments, choice demands a specific, overt comparison, and selection. This technique is used most frequently in educational tests, but can also be used in research instruments, such as the chapter 3 survey.

6. *Scaling* is the most frequently used quantitative self-report methodology. Scaling arranges a condition along a continuum ranging from a zero, null point to some value representing the limit of the condition described.

The continuum is often divided into units indicating equal intervals of greater or less. The essential characteristic of the scale is that it extends the previous methods to numeric weighting, which permits relations to be established (e.g., this mathematics problem is 20% harder than that one). Scaling is among the most frequently employed techniques for research self-reports. More will be said about the scale later.

The significant thing about all these methods is, again, that they require the human to act as a measurement instrument. This means that it is necessary to make certain assumptions (e.g., using the scale as an example: The human can recognize internally generated stimuli; can recognize the verbal labels in a scale and associate these with the internal condition being measured; and the scale accurately describes the range of the respondent's internal stimuli).

These (pragmatic) assumptions are needed to make the measurement possible, but no one really believes that they are completely valid for all or even most individuals. Humans are unique in terms of their internal stimuli, but it is assumed that all respondents are equally capable of measuring those stimuli adequately. All these judgments may require the human to make use of some internal scale that is probably less well articulated than the physical measurement scale.

Every beginning psychology student becomes familiar with the four types of scales: nominal, ordinal, interval, and ratio (Stevens, 1975). Nominal scales merely identify categories. For example, system development has four phases: predesign, detailed design, production, and operations. Nominal scales have no values and cannot be manipulated mathematically. Ordinal scales permit ranking of performance (e.g., rank in a graduating class). Interval scales provide information about amounts of difference, as in a Fahrenheit temperature scale, where 80° is warmer than 60°, and the difference between 80° and 60° is the same as the difference between 60° and 40°. Ratio scales are assumed to possess equal intervals and a true zero point. The 5-point Likert-type scale employed in the chapter 3 survey is an example of a ratio scale.

Purpose

The purpose of the judgmental methods is to elicit information about the subject's perception/cognition of internal (as well as external) conditions. The purpose of the psychophysical methods is to determine thresholds of capability.

Use Situations

The judgmental methods find their greatest use in system-related measurement. Because conceptual research is oriented around the experiment and objective methods, subjective judgmental processes are only occasionally utilized.

The most common application of these methods is in a verbal written or electronic context; most often as part of a questionnaire, only infrequently as part of an interview.

Questions Asked

Although all the judgmental methods relate to internal conditions, each method is designed to answer a particular question:

1. What is the object, phenomenon, event? (classification)
2. How does the object or phenomenon differ from any other? Which is the same, greater, or less than another with which it is compared? (comparison)
3. How much of an object or condition is there? How frequently does a condition occur? (scaling)
4. In what sequence does an event, an activity, occur? (sequencing)
5. Which object, phenomenon, or statement of an event is preferred or is correct? (choice)

Underlying Assumptions

These have already been described. They require the subjects to have a sensory/conceptual frame of reference (schema?), which will enable them to recognize and evaluate internal stimuli in two ways: through description of the subjective condition (feeling) itself, manifested through a scale or classification that represents that condition; or by a deduction from the feeling or condition, expressed as a choice between or a comparison of alternatives. The object of the judgment may be physical, but the judgment itself is the result of analysis of internal stimuli.

The frame of reference utilized by the subject is partially experiential and partially directed by the researcher through characteristics of the measurement instrument. For example, if the scale contains verbal descriptors of internal conditions that can be associated with levels or quantities of those conditions, these enable the subject to translate somewhat inchoate feelings into more overt descriptive categories.

Procedures Employed

The following procedure applies to all self-report methods, but it is here illustrated by reference to the development and administration of a scale:

1. Determine why a particular self-report technique should be employed in a particular measurement.
2. Specify what it is that the researcher wishes to learn from the self-report. The researcher may find it useful to conduct a preliminary interview with potential subjects to find out how much awareness they have of their internal experiences and to discuss the nature of the information they can provide, the precision with which they can report desired data, and so on. The researcher must, however, be careful not to bias the subject.
3. Develop a preliminary form of the reporting form to be administered.
4. Try the scale or other instrument out to determine whether subjects correctly understand instructions and the amount of variability one is likely to encounter.

Advantages

The judgmental methods may be the only effective way of gathering quantitative data about internal conditions.

Disadvantages

The subject is an uncalibrated measurement device and may be less than maximally efficient.

Outputs

These are quantified subjective judgments.

Summary

If researchers wish to quantify subjective responses, then they must use judgmental methods.

HUMAN PERFORMANCE SYSTEM MODELS

Description

The reason that human performance system models are included in a discussion of subjective measurement is that the model is a human construction, hence subjective. These models are potential vehicles to study system variables when it is difficult to manipulate the actual systems. It is often not feasible to study molar variables like system complexity in actual systems like process control plants, whose size makes it difficult to manipulate them as total systems. If a symbolic system were actually equivalent to the real one, then the latter's variables could be included in the model system and results that could apply to the actual one could be derived.

Although HF is about human–machine systems, their measurement in HF is usually not of the systems as a whole, but of individual items of equipment, like icons or menus. The assumption is that if knowledge of system components is additive and can be combined, then the combination would equate to actual system knowledge. Needless to say, this assumption is another pragmatic one, and probably somewhat fallacious. System models could remedy this. Such models could be used in system development to compare design alternatives, to find the most effective one, or to test a design concept. For example, the U.S. Navy is presently working to develop a destroyer (DD21) with a complement of only 95 sailors, as compared with present destroyer complements of 300 or more. If it was possible to symbolically represent the projected DD21 system, equipment, and personnel, in advance of their development, it would be possible to determine in advance if such a ship can, in fact, be operated with only 95 sailors. As it is, planning on the project proceeds under the untested 95-sailors assumption (previous Navy efforts to reduce manning have failed in the past).

The sticking point is, of course, that the model must function like the actual system or class of systems. This requirement must be verified by exercising both the model and the actual system, and then comparing their performance. This may seem like a difficult hurdle to overcome, but it will not be necessary to perform such comparisons in every case, if enough confidence can be built up on a few test prototypes.

There are many types of human performance models. Some deal only with individual human functions, such as strength or vision; others describe the interaction of an operator with an equipment. Still others attempt to reproduce an entire system as operated by its personnel. The latter models, with which this discussion is concerned, fall into two general categories described as reductionistic (task network) and functional, or as Laughery,

Corker, Campbell, and Dahn (2003) called it, "first principle" models. Reductionistic models are structured in terms of human/system task sequences. Functional models are built around goals and principles represented in behavior functions like perception, motor response, and cognition. The two types of models interact, because humans are both goal-seeking and task-performing. The following discussion is concerned with task network models.

A subspecies of the general performance model is what can be called the *human error model,* because its emphasis is on the prediction of the human error probability (HEP) of a system.

Like other aspects of HF activity, human performance system models depend on computer technology. Because of this, in a future of anticipated expansion of this technology, the symbolic model will presumably be so improved that it may well offer the best hope of researching large, complex systems.

It is, of course, impossible in this short section to deal fully with the extensive literature on these models. That literature has been summarized in Salvendy (1997). The most interesting models are organized around task networks, because human–machine systems are also organized in terms of tasks and missions to be performed in order to achieve specified goals.

This means that a prior task analysis (or at least a task sequence) is critical to the development of the model. The task, in general terms, is the operation by personnel of a type of equipment, such as a valve or workstation. Archival data describing the performance of personnel in operating a class of these equipments (e.g., switch, work station) are input by the developer into the model; operator performance of tasks early in the sequence modifies the performance of tasks occurring later. The model scenario progresses in simulated mission time until the terminal task, which leads to mission accomplishment. All the preceding is incorporated in a computer software program, which is organized in terms of rules for exercising the models.

The structure of a computerized network, as described by Laughery and Corker (1997) consists of the following: (a) task number and name for identification; (b) task performance times and error probabilities sampled from a distribution that is predetermined (e.g., normal or exponential); (c) mean time or average performance time of the task, expressed in a number, an equation, or algorithm; (d) the standard deviation of the task performance times; (e) a release condition, which is a condition that, if not satisfied, will hold up task execution until the condition is satisfied (e.g., an operator must be available before the task can begin); (f) a beginning effect, a field, which defines how the system will change once the task begins;

(g) a launch effect, which is the same as a beginning effect, but used to launch high-resolution animation of the task; (h) an ending effect, which defines how the system changes when the task is completed.

In a task network model, several tasks may commence at the completion of a previous task. This means that the operator must decide which of a number of potential courses of action should be followed:

1. *Probabilistic*—task selected on the basis of random draw weighted by the probabilistic branch value.
2. *Tactical*—beginning one of a number of tasks based on their highest value.
3. *Multiple*—several tasks beginning at the conclusion of the previous task.

Obviously, the model functions on the basis of rules, just as an actual system functions. The rules can be expressed in various ways (e.g., numbers, algorithms, equations, etc.).

Although the aforementioned discussion describes the human performance model in general, each model must be adapted to represent the particular system or type of system it simulates. The specific model is built up using a general operating system like SAINT or MicroSAINT as a foundation, via a computer point-and-click drawing palette to represent the specific tasks performed by the actual system.

The one element, which the preceding system model description has not emphasized, is the inclusion of error data. It is impossible to think of actual human–system performance without the possibility of error. Any valid system model must incorporate error probabilities based on a distribution of such probabilities. This suggests that the outputs of any human error model (or rather the data collected to support such a model) should serve as an input to the system model.

The human error prediction model is also organized around tasks. Error data (time is also important, but not as important to this type of model as error) are input to each task, as before, and the error rates of these tasks are then combined according to specified rules to secure an error probability for the entire system.

As has been pointed out earlier, the basic metric of the human error model is human error probability (HEP), which is defined by the following equation:

$$\text{HEP} = p = \frac{\text{number of operations in which one or more errors occurred}}{\text{total number of operations}}$$

Purely as a hypothetical example, if a task were to be performed 100 times (total number of operations) in controlling a machine, and three errors were made during these,

$$\text{HEP} = \frac{3}{100} = .03.$$

The point is that any mission is composed of many tasks whose individual error probabilities must be combined in some way. A number of different combinatorial techniques have been proposed (Swain, 1989), the most frequently used being technique for human error rate prediction (THERP; Swain & Guttman, 1983).

THERP, which is based on task analysis, models events as a sequence of binary decision branches in an event tree. At each node in the tree a task is performed correctly or incorrectly. Except for the first branch, the probabilities assigned to all tree limbs are conditional probabilities (i.e., based on a previous task with a previous probability). By accumulating probabilities, the researcher arrives finally at a cumulative probability, which describes terminal (end of sequence/mission) performance. The product of such cumulative probabilities can be used in two ways: to predict the overall probability of correctly operating an equipment or system (1 minus the cumulative error rate), and to select equipment components or operating sequences on the basis of the lowest error rate predicted for them.

This combinatorial process is far more complex than the description implies. Unless greater success has been achieved than in the past, its predictive power is less than desired.

A summary of the outstanding characteristics of the two models described in this section makes it is obvious that there is much similarity between the two. Both require input data of a behavioral nature (task/error data); the system model also requires physical (equipment) data, such as radar capabilities. Both function in terms of rules for the way personnel performance is organized. Both require the combination of lower level (task) performance values to derive higher level (subsystem and system) performance values. Both provide quantitative predictions of how well personnel will function in performing an action.

The few differences between the two are, however, important. The major difference is that in the system model the specific error/time data applied to the task are taken from a distribution of those data; this permits individual performance variations to occur. In the human error model, the error data are not taken from a distribution, but from a fixed equation, which means that random variations in performance are not incorporated. The human error model is exercised only once; the human performance model usually involves several repetitions, resulting in a distribution of terminal values. This makes the system model more dynamic than the human error model. The human error model, like the system model, takes into account variations resulting from fatigue or workload, but not as effectively as does the system model.

There has been little interaction between those researching the two models. This may be because they derive from two different orientations. The human error model stemmed from an attempt on the part of HF professionals in the 1950s to find a behavioral analogue to methods used by reliability engineers to predict equipment reliability (see Meister, 1999). The first cohort of HF professionals working in industry, and, exposed to reliability engineering, were impressed by the latter's predictive capability. They saw in human error an analogue of equipment failure and adopted reliability methodology as a model for their own prediction efforts. As it turned out, the same multiplicative model used to predict equipment reliability ($a \times b \times c \ldots N$, where a, b, c, etc. were the reliabilities of individual equipment components) was too simple to predict human performance, because the model could not include human flexibility.

The system models were developed out of a psychological tradition to which the reliability orientation was completely foreign. Nonetheless, the assumptions underlying the human error model are perfectly applicable to the system model.

Purpose

The purpose of the models is to describe/reproduce in symbolic form the functioning of humans and equipment in order to predict the performance of the system they compose.

Use Situations

The model has certain potential uses to which it can be put, including:

1. To describe how personnel represented in these models perform their functions.

2. To predict the adequacy of humans operating the system in the performance of specified missions. That adequacy may be described in various ways: whether or not the mission goal is achieved; the time it takes to achieve that goal; in the case of the human error models, to predict the error rates manifested by personnel in performing tasks.

3. To compare system performance under varying configurations involving differences in such parameters as type of equipment operated, number of personnel, nature of the tasks performed, and task and system organization. These comparisons are analogous (symbolically) to the performance of experiments, if these were performed on actual systems. The purpose of such comparisons is to determine which system configuration (design alternative) is most effective.

4. To perform research on variables that generally affect personnel and system performance, apart from studies on specific systems. For example, Laughery and Corker (1997) presented a case study of crew mental workload and suggested the following possible comparisons: modifying task times and changes in the time required to access a new display; modifying task times and accuracy based on changes in the content and format of displays; changing task sequences, eliminating tasks, and/or the addition of tasks based on changes in system procedures; changing task allocations.

The previous functions can be performed in two ways. The comparison of alternative design configurations could be performed in parallel with (as part of) or preliminary to the development of an actual system (e.g., like the construction of a physical aircraft simulator, which is used during aircraft development to try out solutions to various design problems).

The models could also be used to research system variables (e.g., what is the effect of system complexity on personnel performance), independent of specific system developments.

Questions Asked

Two types of questions can be asked of a system model: those asked before exercising the model and those asked afterward. The questions asked before model performance include: (a) Has the model been adequately verified, so that the researcher can have confidence in its outputs? (b) Are the input data (both human and equipment) adequate to exercise the model as it was designed to be exercised?

The questions for which the model outputs will provide answers are:

1. How well does the symbolic system and its personnel perform in terms of specific measures of effectiveness (e.g., time to accomplish the mission goals and any accompanying error rate)?
2. What factors (e.g., equipment and task changes, organization of procedures) affect system personnel performance, and in what way?
3. How do alternative system configurations change system and personnel performance?

Assumptions

The central assumption of every model is that the rules and data by which the model functions are the same or very similar to those with which the real-world system, which it is modeling, performs. In other words, the model correctly simulates the real-world system, including all the essential elements of the real-world system.

More specific assumptions are made by the human error model (Reason, 1987) that apply just as much to the human inputs of the system model. These assumptions also describe psychological processes that underlie all human performance:

1. Cognitive control processes operate at various levels and exert their influence simultaneously over widely differing time spans.
2. These higher levels function over both long and short time spans, and a wide range of circumstances. They are used to set goals, select the means to achieve these goals, and to monitor progress toward these objectives.
3. Higher level agencies govern the order in which lower level processors are brought into play.
4. A substantial part of cognitive activity is governed by lower level processors (schemata, scripts, frames, heuristics, rules) capable of independent function. These lower level processors operate over brief time spans in response to very specific data sets.
5. The successful repetition of any human activity results in the gradual devolution of control from higher to lower levels; this is represented in the familiar learning process.

Procedures

Assuming that the system will be represented by a task network model, the following is the general procedure to be employed in developing the model:

1. Study the task analysis that has been performed of the real world system.
2. Decompose the system task analysis into the mission, submission, functions, subfunctions, tasks, and subtasks to the level of detail required to depict the system.
3. Develop the relations (network) between and among tasks.
4. Select a computer simulation engine like SAINT or MicroSAINT as the basic set of building blocks for the model. The method of applying

SAINT or MicroSAINT is described by its reference description (Micro Analysis and Design, 1993).

5. Determine what input data are required to animate the model. This will include the listing of mission, functions and tasks, their interactions, and in particular the conditions under which they function (the rules imposed on the model) and the distribution of times and errors by which system elements function.

6. Input 5 into the model. The procedures to be followed are described in references like Micro Analysis (1993). See Laughery and Corker (1997) for concrete examples.

7. Try the model out, noting its outputs over time, and comparing these with performance data from the system or system type it is modeling. Model development will almost certainly require progressive refinements until there is a reasonable match between the symbolic model and any real system.

The general procedure for developing the human error model (using THERP as an example) is to build a fault tree of tasks based on binary decision branches, where the decision is whether or not the operator performs that task correctly. The drawing of the branches of the tree is dependent on task analysis. The error probabilities assigned to the decision branches are presumably based on archival data, modified by assumptions and principles founded on experience, logic, and intuition. For example, THERP divides dependency among task elements into five discrete levels, and the original error probability is modified by mathematical rules, which take the different degrees of dependency into account. Much greater detail is to be found in Park (1997) and Swain and Guttman (1983).

Advantages

If the models are valid representations of actual systems, then it will be possible to perform developmental and research studies on the model systems without having to manipulate the actual systems. The model also permits multiple repetitions of system operations, which is impossible with actual systems because of operational constraints. All of this will make system studies easier to perform. In addition, the model permits quantitative prediction of human performance, which present research does not permit.

Disadvantages

Most models are only partially validated on an empirical level. The model is also heavily dependent on data inputs (i.e., time and error distributions), which are often not available or are incomplete or insufficiently detailed.

Davies (1998, abstracted in Meister & Enderwick, 2001, p. 79) listed disadvantages that present models possess.

Outputs

These are quantitative, in some cases probabilistic, estimates of personnel and system efficiency. Outputs are keyed directly to the questions asked and the uses of the model.

Summary

It is difficult to make definitive statements about models, because surveys of actual model use are lacking. It is quite possible, however, that present modeling activity is primarily a research function; that such models, because of the cost in time and effort to develop and use them, are used mostly by research agencies. Nonetheless, the use of models in the future is strongly indicated, because they have great potential as measurement instruments for complex systems.

New system models continue to be developed. For example, although SAINT and MicroSAINT are probably the most well known of system engines, they represent a modeling technology that is 20 years old. Many newer models are being developed at, among other facilities, the Ohio State University by people like Dr. David Woods and his students.

FORENSIC INVESTIGATION

Description

Forensic investigation may be thought of as a distinct methodology, which makes use of a basket of other measurement methods such as observation, interview, available HF data, problem solving, even in some cases experimentation. What makes forensic investigation interesting is that it is the most marked real-world application of HF methodology and data, with the exception of system development.

Forensics is a form of detective story with legal overtones. An injury to a plaintiff has presumably been committed by a company or an individual, and the resultant activity is directed at determining who, if anyone, is at fault. The injury is human related (e.g., the plaintiff in the suit has fallen down a flight of restaurant stairs and alleges the restaurant was at fault). The cause of the injury is alleged to be some behavioral factor (e.g., the restaurant stairs were not adequately lighted so the patron could not see, or too steep). The question becomes one of fact to be presented in a court of law.

The problem is presented as a question: Were the environmental conditions under which the injury occurred within the plaintiff's threshold capabilities, so that it would be expected that the individual should have performed adequately, or the reverse?

The answer to this question for the HF specialist—whose role is to serve as a consultant or expert witness—will be derived from application of one or more of the methods described in previous chapters. For example, an experiment could be performed, with the situation in question (stairs, lighting) being utilized or replicated. Several subjects with normal eyesight and motor capability could be asked to descend the stairs under various lighting conditions. The plaintiff and other restaurant patrons could be interviewed with regard to how difficult it was to descend the stairs. The stairs themselves could be observed by the specialist and objective measurements of visibility made of the lighting and the stair angle. The physical condition of the stair carpets could be examined if the plaintiff had stumbled. These measurements could then be compared with lighting standards, prescribed stair angles, and the minimal coefficient of friction of the stair carpets.

What is at question here from the standpoint of the HF discipline is the adequacy of HF standards and data, methods of collecting new data, and their interpretation and application to a particular situation.

Purpose

To determine whether a specific human performance (e.g., a fall) occurred as a result of exposing the humans to conditions outside of their (threshold) capability to respond adequately.

Questions Asked

The following questions should be answered:

1. What is the human threshold capability for a particular behavioral function?
2. Was this threshold capability exceeded and what were the causal factors?
3. Did exceeding the threshold capability produce the injury of which the plaintiff complains?

Use Situation

Forensic investigations are performed in the context of a real-world legal situation. Problem solving is the essence of the HF methodology, as is the collection of data from the operational environment. Of particular importance is the question of the ability of archival HF data to describe validly a *specific* (rather than a general research) performance. The emphasis here is precisely on the applicability of HF methods and data to a nonresearch situation.

Assumptions

HF forensic assumptions include the following:

1. The available HF data and measurement methods are adequate to describe the capabilities and limitations of specific human behaviors.
2. The HF data in reference sources and any new data produced as a part of the investigation are relevant to the human performance under inquiry.
3. These data can be used to explain the human's performance in that situation.

Procedures

1. The forensic investigator begins by analyzing the questions posed by the legal situation. This analysis helps to determine the parameters of the problem and the specific questions to be answered by the HF investigation.
2. The performance situation (the one in which the injury occurred) is recreated either symbolically or physically.
3. The parameters of the HF data needed (e.g., lighting standards, visual resolution, stairway parameters) are determined.
4. The investigator selects one or more appropriate methods to gather empirical data applicable to the problem.
5. The relevant HF data are collected and analyzed. The investigator determines whether these data are adequate to support the case.
6. The data collected, the methods employed, the conclusions reached, are documented; this is followed by interviews with counsel and testimony in court (if it goes that far; many cases do not).

Advantages

One of the things that forensic investigation can do for the HF discipline is to indicate the adequacy or inadequacy of the HF data gathered during previous research to describe a particular functional area (e.g., visual perception). The legal setting creates a stress situation whose resolution reflects and tests (at least partially) the adequacy of HF methods and data.

Disadvantages

It may turn out that available HF data and methodology do not adequately support what is required by the forensic investigator. HF methods and data in forensics, like the same measurement methods and data in design, are taken out of the "scientific" but general context and are now in a "use" context, which is a more rigorous challenge. The inability to meet this challenge (if that is indeed the case) may be disturbing to the HF professional.

Outputs

The major output of the specific investigation is an answer to the question that originated the investigation: Were the technological constraints imposed on the human sufficient to cause the injury complained of?

Summary

It is unlikely that HF professionals will look at forensic investigation as an applied test of HF measurement adequacy—but they should. The application of HF data to design is another such test, but the forensic test is much more immediate and obvious. The forensic problem, like the design problem, tests the capacity of the discipline to supply required procedures and, particularly, data.

CRITICAL INCIDENT TECHNIQUE

Description

Critical incident technique (CIT), developed by Flanagan (1954), has achieved great success in terms of frequency of usage. One reason for this is that there are the affinities between the technique and the storytelling that has been a part of man's heritage since speech was developed.

The basic assumption behind CIT is that an extreme example of a phenomenon illustrates/highlights characteristics of that phenomenon and makes it more understandable. The incident (story) in the CIT often presents a problem that must be solved. The incident, when reported adequately, expresses the major aspects of a situation, something accomplished by all good storytelling.

The incident is, by its nature, one of a kind; but it may also have similarities to other incidents, which means that a pattern may be discovered. CIT can be thought of as a method for analyzing (decomposing) a situation to determine, for example, what factors caused the situation, the effect of the incident on the respondent, and what resolved the problem represented by the incident. If there are several incidents for a given situation, these could be analyzed to find their common elements. Then, based on that analysis, another study could be conducted. Obviously, a problem is suspected before the CIT is activated.

Researchers elicit the incident by asking for it, so it is necessary for them to specify the topic of interest. For example, they might ask about any human performance problems that arise while operating a nuclear reactor control room. The general parameters of the problem for which they wish to receive an incident must be specified.

Purpose

The purpose is to describe the essential characteristics of a problem situation and to assist in the analysis of that problem.

Use Situation

The critical incident, like the interview, can be elicited almost anywhere. If the respondent who supplies the incident is asked further questions about the incident, CIT transforms into an interview.

Assumptions

The basic assumption is that the extreme characteristics of a situation reveal the essential elements of that situation.

Procedure

The following procedures should be observed:

1. Explain to the respondent the type of problem situation to be described. For example, ask for a description of any critical (outstanding) incident that may have occurred during the aircraft landing process.
2. Ask the respondent to try to remember one or more incidents that illustrate the characteristics of the problem.
3. Record the incident in the respondent's own words, using a recorder.

4. Analyze the incident in terms of the nature and severity of the problem, the conditions that apparently produced the incident, and the short- and long-term effects on the subject and/or on the system. Determine how the problem in the incident was resolved.

Advantages

The incident has the advantage of being quite "natural," in the sense that it is simply storytelling, which presumably anyone can relate.

Disadvantages

The validity of the incident is dependent on the intelligence and memory of the respondent, as well as the latter's lack of bias. The incident also does not represent *all* instances of an activity, and one requires several subjects, because what may have impressed one respondent may have completely passed another by. The incident is almost always entirely qualitative.

Outputs

The critical incident stands alone; it is merely illustrative. It supports further investigation and is used with other methods. One can perform a content analysis of the incidents, but CIT does not itself produce outputs.

Summary

CIT has a very limited use. It enables the respondents to express themselves more clearly than if there were no incident to report, and than if they relied solely on an interview. Although purely descriptive, it tends to highlight the outstanding characteristics of a situation and thereby serves as a guideline to the researcher. CIT is a preliminary to other research efforts; a researcher would certainly not collect incidents and then do nothing further with them.

Chapter 7

Measures

The development and selection of measures is far more complex than the average HF professional thinks. Measures, like measurement as a whole, are supported by a conceptual structure of which the reader should be aware.

In fact, of all the measurement processes described so far, the development of new or application of old measures to describe a particular performance is the most demanding. At least part of the difficulty arises from the fact that human performance is a physical phenomenon, and any measure of that performance (e.g., error) is only symbolic or conceptual. Nevertheless, that symbolic index is the only one available to represent the performance.

Human performance measures in conceptual studies indicate only personnel *capability* to perform, because, unless there is an interaction with physical equipment, the human performance in the conceptual study is only half realized. Human performance measures in system studies are more veridical representations of actual human performance, because their behavioral aspect is changed slightly by their interaction with an equipment. For example, the human's ability to detect using a radar is modified by the brightness of the radar's pips. System measures are, therefore, no longer behavioral measures alone, once they are combined with physical measures. It is this combination of human with physical measures that may make the meaning of a purely behavioral measure (i.e., excluding equipment/system aspects) obscure. (But how about a conceptual study in which subjects are required to operate equipment? Such a study, even if it is performed in a laboratory and the equipment is only a simulation, is almost a system study. However, unless the task/equipment simulation is close to the "real thing," the study will lack something.)

The purpose of the behavioral measure is to describe and quantify human performance. Table 7.1 lists generic behavioral and behaviorally related measures. These measures include the following dimensions: time, accuracy, frequency, accomplishment, consumption of resources, physiological and behavioral descriptors, and categorization of behaviors by oneself and others. These general categories must be made more precise by application to the specific function, task, and system situation being measured.

It might seem that all that must be done to select an appropriate measure is to examine Table 7.1 and other compilations, like Gawron (2000), to find an appropriate category and choose. This is incorrect. The process of selecting an appropriate measure is an analytic one, more complex than many professionals may realize. Part of the routine of customary measurement practices is to shorten the measure selection process by collapsing its stages into each other. Researchers may accept obvious measures without working through their relations to other aspects of performance, more specifically functions, tasks, and the mission. If they do, however, the effectiveness of their measures will suffer.

The relations referred to occur only within a system and within system-related measurement. For conceptual research, in which a system and its context do not exist, or exist only as a possible reference, there can be no relation between the measure and the system. In conceptual research the only relation that exists is between the performance of the individual subjects and the *function* of the task they are performing. (This applies, of course, only if the task is synthetic, i.e., nonoperational.) Lacking a system mission, conceptual research can supply data relevant only to the subject's capability and can say nothing about any relations of the subject's measure to higher order performance levels.

The measure is inextricably linked to performance outputs; it describes how these outputs function. In any but the simplest systems there are many potential outputs, and this may pose a problem of choice for the researcher. Because conceptual research lacks a system, with its own system outputs as well as those of its personnel, the number of performance outputs in this kind of research is somewhat restricted.

The only outputs produced in a conceptual study are those that describe individual subject performance, and so the accuracy and time measures so typical of conceptual experiments appear to be sufficient. In real life, however, the human works as part of a system and so the operator's performance must be related to system aspects.

Certain criteria for selecting measures can be specified. The measure should be highly related to the output or product of the performance being measured (in the case of humans, accuracy of task performance; with systems, system success or failure); objective; quantitative; unobtrusive; easy to collect; require no specialized data collection techniques; not excessively

TABLE 7.1
Generic Performance Measures

Time

1. Reaction time, i.e., time to:
   a. perceive event
   b. initiate movement
   c. initiate correction
   d. initiate activity following completion of prior activity
   e. detect trend of multiple related events
2. Time to complete an activity already in process; i.e., time to:
   a. identify stimulus (discrimination time)
   b. complete message, decision, control adjustment
   c. reach criterion value
3. Overall (duration) time
   a. time spent in activity
   b. percent time on target (tracking)
4. Time-sharing among events

Accuracy

1. Correctness of observation; i.e., accuracy in:
   a. identifying stimuli internal to self and system
   b. identifying stimuli external to self and system
   c. estimating distance, direction, speed, time
   d. detection of stimulus change over time
   e. detection of trend based on multiple related events
   f. recognition: signal in noise
   g. recognition: out-of-tolerance condition
2. Response-output correctness; i.e., accuracy in:
   a. control positioning or tool usage
   b. reading displays
   c. symbol usage, decision making, and computing
   d. response selection among alternatives
   e. serial response
   f. tracking
   g. communicating
3. Error characteristics
   a. amplitude
   b. frequency
   c. content
   d. change over time
   e. type
      (1) reversible/nonreversible
      (2) relevant/irrelevant
      (3) effects: significant/nonsignificant

Frequency of Occurrence

1. Number of responses per unit, activity, or interval
   a. control and manipulation responses
   b. communications
   c. personnel interactions
   d. diagnostic checks

(Continued)

TABLE 7.1
(Continued)

2. Number of performance consequences per activity, unit, or interval
   a. number of errors (of any type)
   b. number of out-of-tolerance conditions
3. Number of observing or data gathering responses
   a. observations of internal/external events/phenomena
   b. verbal or written reports
   c. requests for information

Amount Achieved or Accomplished

1. Response magnitude or quantity achieved
   a. degree of individual/system success
   b. percentage of activities accomplished
   c. achieved numerical reliability
   d. measures of achieved maintainability
   e. equipment failure rate (mean time between failure)
   f. cumulative response output
   g. proficiency test scores (written)
2. Magnitude achieved
   a. terminal or steady-state value (e.g., temperature high point)
   b. changing value or rate (e.g., degrees change per hour)

Consumption or Quantity Used

1. Resources consumed per activity
   a. fuel/energy conservation
   b. units consumed in activity accomplishment
2. Resources consumed by time
   a. rate of consumption

Physiological and Behavioral State

1. Operator/crew condition
   a. physiological (e.g., stress level)
   b. behavioral (e.g., workload)

Behavior Categorization by Self and Others

1. Judgment of performance
   a. rating of operator/crew performance adequacy
   b. rating of task or mission segment performance adequacy
   c. estimation of amount (degree) of behavior displayed (e.g., scaling)
   d. analysis of operator/crew behavior characteristics
   e. estimation of system qualities (e.g., Cooper & Harper, 1969)
   f. agreement/disagreement with concepts (e.g., chap. 3)
   g. preference for alternatives
   h. estimation of behavior relevancy
      (1) omission of relevant behavior
      (2) occurrence of nonrelevant behavior
2. Analysis
   a. stimulus recognition
   b. classification/comparison of stimuli
   c. decision making
3. Capability measures (e.g., aptitude scores)
4. Subjective reports
   a. interview (verbal/questionnaire content analysis)
   b. self-report (verbal/written) of internal status/external events
   c. peer, self, or supervisor ratings

Note. Adapted from Smode, Gruber, & Ely (1962) and Meister (1985).

molecular, which usually requires instrumentation; and cost as little as possible. The first criterion is the most important and admits of no mitigation; measures that do not suggest how human–system performance has occurred are irrelevant. For example, if researchers are testing mathematics knowledge, measurement of the subject's eye movements during the test will usually say nothing about that knowledge. The other criteria are desirable, but can be modified if the nature of the measurement situation demands it.

One of the important aspects of the measure is that it forces the researcher to select certain aspects of the performance being measured as more important than others. (This is also linked to hypothesis development.) There are some facets of performance that researchers might not wish to measure, because of their relative unimportance. Nontask-related performances, such as the individual's body language, are usually not measured, because these largely idiosyncratic aspects are not task and goal driven (of course, body language may be important in certain specialized games, like poker, in which they are goal related).

A measure has the following other uses:

1. Measures describe what people actually do as compared with what they are supposed to do. The task as described in a manual suggests no subject variability or performance inadequacies.

2. They enable researchers to measure general (nontask specific) capability, as in visual or intelligence tests. When performing tasks, measures tell how well subjects perform these.

3. They serve as a means of quantitatively relating and differentiating treatment conditions.

4. In addition to experimental treatment conditions, measures help to suggest those factors that influence performance. Thus, they also serve an indirect diagnostic function. For example, a higher incidence of error linked to a particular function or mission phase may suggest that something connected with that function or mission phase is influencing performance negatively.

5. Theoretically, behavioral measures permit us to relate human to system performance. More on this later.

The preceding discussion has deliberately not dealt with statistics (e.g., *t* test, ANOVA), except to note that measures are used to differentiate performance differences between treatment conditions, and of course significance level statistics are involved in that process. Discussion of statistics requires much more extensive treatment than this text permits.

THE DEFINITION OF A MEASURE

A measure has three aspects. First is the *metric* inherent in the measure. The metric incorporates two quantitative dimensions: number and time. This enables the researcher to apply the measure quantitatively, by performing basic human functions like counting.

Number and time are fundamental (generic) measures because they apply to all performance, although the specifics of the measurement situation determine whether each will be used. Number is important because performance outputs usually occur in multiples; time is important because all performance occurs over time—from the initial stimulus to the final response.

Examination of Table 7.1 reveals that time and, even more, error are the heart of the measure, because they are capable of many variations expressed in terms of the various functions to which they refer (e.g., stimulus identification, reading displays).

To count (e.g., errors) is to perform a behavioral function; that performance becomes a *method* permitting the metric to be applied. The thing counted can be anything: eye-blinks, errors, items retained in memory. The "anything" requires analysis of functions, tasks, or performance of a specific type. For example, if researchers perform a study involving memory, the nature of the memory function (retention) suggests a method for utilizing the number metric: number of memorized items retained after a specified interval. To secure such a number, the measurement of retention requires that subjects be tested at least twice: after initial learning, and following an interval (which will be filled by performance of other functions and tasks).

The metric, together with a method of applying the metric, becomes the measure. A measure cannot exist without a method of applying the metric. The method depends on human capabilities. The use of number requires the human capability to count objects, to differentiate among objects of different types, and to recognize similarities among these objects. Because the human has great difficulty measuring time subjectively, instrumentation—a timer—must be used.

The capabilities humans possess determine the measures that can be applied. Humans can analyze their own status and perceive and report external status (hence self-report techniques); they can perform and their performance can be measured with or without a standard; they can perform individually or in relation to a system (hence conceptual and system-related studies); they can indicate a preference or choose among alternatives (hence multiple choice techniques); they can perform more effectively in one configuration or situation than another (hence experimental treatment conditions); they can analyze relations (hence researchers can infer the effect of variables on their performance); they can combine and interrelate objects (hence the development of predictive equations).

All the behavioral functions available to all people are involved in the measurement methodology: stimulus detection and recognition, stimulus analysis, comparison of stimuli, choice among stimuli alternatives, decision making, and so on. These basic human capabilities, which function ordinarily as part of all human activities, can be transformed into measurement methods. For example, the capability to verbalize is turned into such self-report techniques as the interview/questionnaire; the human's ability to evaluate is turned into a rating or a scale; the ability to judge is turned into a decision by the subject. These human capabilities are utilized most often in subjective measures, but the most popular objective measure—error—depends on the researcher's ability to recognize an error.

The performer (the subject) also determines the kind of measure employed. If an investigator measures performance using a number (or its most common variation, error) without directly involving the subject in what is measured, the result is an objective behavioral measure. If the method chosen requires direct interrogation of the subject, then the measures involved are subjective.

Number and time are used in both objective and subjective performance, although the use of time as a subjective measure is very infrequent, because many tasks are not time dependent, although almost all are error dependent. Subjective measurement affords many more methods of metric application. As was pointed out in chapter 6, researchers can make use of the subject's functional capabilities (motor, perceptual, cognitive) to expand ways of applying the number metric. For example, the human capability to compare stimuli has been used to develop the paired comparison methodology. Paired comparison is largely subjective, because objective comparisons are not possible except in experimentation. However, relatively few experimental methods permit comparison of multiple treatments.

Because every measure is dependent on the nature of the function/task being measured, a measure cannot be defined without knowing the function/task to which the measure refers. All observed performance is meaningless unless that performance can be related to a function and task and their purpose.

The specific measure is therefore determined by the nature of the function/task being described by the measurement. For example, the metric for measuring detection or memorization performance may in both cases be error, but the kind of error counted depends on the individual function (detection or memorization).

Any performance is the translation of the function/task into action. That function/task is described verbally in procedures and manuals, but only statically, as an abstraction; unless it is seen in action, it remains an abstraction. Moreover, there is no way to tell how the function/task is actually performed unless the human is required to implement it. The verbal description represents an ideal. That description assumes no performance variability because, as described verbally in the manual, there is only one way to perform that function/task. Because the function/task is performed by a human, and humans are inherently variable, there will always be some variation between the abstract verbal description and its actual performance. (This normal variability is not, however, the same as that determined by performance-shaping variables.) That is the rationale for all performance measurement. The measure is the means of describing the difference between static (hence ideal) and dynamic (real).

Subjective measures are more open-ended (possess more alternatives) than are objective ones, because the former make use of the human as an uncalibrated measurement device. The self-report of the subject can be turned into ratings, scales, and multiple choice selections. Objective measures make use primarily of the metrics of time and error (in all their variations). All measures, including purely qualitative ones like opinions, can be quantified, although the latter may have to be transformed into a related metric.

Frequency statistics (e.g., mean, median, standard deviation) provide useful information to the researcher, but significance level statistics (the great majority of abstruse statistics like ANOVA) do not tell the researcher much except that treatment differences were not the result of chance effects. The treatment conditions themselves represent the variables hypothesized previously by the researcher.

It may appear as if once the measure is selected, there is no need to be concerned with it further. This is untrue. It is not too difficult to select a measure, but it is also necessary to make sense of that measure (what it means). Most measures are not immediately understandable in terms of the specific functions and tasks being measured and require further analysis. For example, unless time is critical to adequate performance of a task, time measures will mean little. Errors may be indicative of performance quality, but only if the errors are relevant (i.e., have a significant effect not only on the performance of the individual but also on system performance). Skill is an important factor; in highly skilled performance, there may be so few errors that data collectors have difficulty in observing them.

Table 7.1 describes only generic measures, from which more specific ones can be derived by relating them to the specific task being measured. Generic measures can also be transformed into methods by taking advantage of the human's innate capabilities. For example, it has been pointed out that the human's capability of making comparisons among objects and phenomena can be formalized into the paired comparison method (Thurstone, 1927). The ability to rate and rank these objects and phenomena can be transformed into formal methods of rating and ranking by giving subjects specific instructions. For example, the judgment capability can become the 5-point Likert scale.

Methods of measuring performance can also be based on theoretical concepts. The outstanding example of this is the use of secondary tasks to measure workload; the secondary task is based on the concept of limited attentional resources; the second task imposes a demand on those resources, and the human's response to that demand "measures" response to the primary task, which, when reduced, is called the effect of workload. If the theory or concept is wrong or inadequate in some way, what is measured will bear little relation to reality.

One can theoretically measure anything and everything, but does that measure make any sense as it relates to the researcher's conceptual orientation to the study questions? More important, will the measure illustrate or at least be related to some important aspect of reality? The measurement situation is not or need not be reality as it is seen in system functioning in the operational environment. That situation includes many performances that are trivial in relation to some reality that the researcher thinks is important. It is possible, for example, to measure the number of operators' yawns as a measure of alertness, but is this important relative to, say, the function of detecting enemy targets?

Measurement itself can be in error. Researchers can measure the wrong behaviors; they can measure those behaviors incorrectly; what they measure may not answer the study questions, or do so only partially. The researchers' first task (in which they often fail) is to understand that the measurement situation is not necessarily reality.

If researchers ask a question, either of the study topic or as a question of the operator subject, there is some conceptual orientation underlying the question. They do not need to know what that orientation is to ask the question or to measure performance, but the more they understand the conceptual foundation of their questions and measures, the more likely they are to get meaningful measurement results.

MEASURE CHARACTERISTICS

It is useful to consider measures as they have been compiled. Gawron (2000) compiled measures that have been published, which raises the interesting point: Are there measures that have been developed and used, but that have not been published and, if so, would these be different from those that have been published? It is likely that many measures employed by professionals were created for special purposes and have never been published, were never validated, may not even have been pretested, and had only a one-time use. Such measures, however, are based on the generic measures of Table 7.1.

Error/time measures are ordinarily applied by researchers to a specific test without any effort at validation of those measures for that test, perhaps because the measures are assumed to be inherently valid as innate dimensions of performance. It has already been pointed out, however, that as far as error is concerned, the error measure applies mostly to highly proceduralized tasks in which any deviation from the procedure is automatically defined as an error. In contrast, with tasks that have built-in variability or options, such as motor tracking or maintenance tasks, consideration must be given to the inherent variability of the human manifested by tremor or alternative strategies; only when normal variability exceeds some minimum can it be considered an error. That error is assumed to be the result of some systematic factor that is responsible for the more than normal variability (Chapanis, Garner, & Morgan, 1949). This factor may not be immediately obvious and may require analysis.

Many subjective measures and methods are not standardized. Interviews designed to uncover performance details, as in posttest briefings, and interviews designed to reveal attitudes are usually developed by researchers on their own, using guides described in textbooks (Meister, 1985). The same is true of questionnaires.

Scales such as those described in chapter 3 are often developed by the researcher for the purpose of a single study. The use of a Likert-type scale format makes it comparatively easy to conceptualize a complex concept as a linear scale and to quantify the concept by dividing the scale into equal(?) halves and quarters, assigning verbally equivalent labels to them (see chap. 3 for illustrations). The result is a scale, which, however, may not actually accord with the subject's own internal scale, when the scale is used to measure highly complex phenomena such as concepts, workload, stress, or fatigue.

All this nonstandardized, nonvalidated development and use of subjective measures may make the measurement specialist uneasy, because it suggests that much data are collected without sufficient care. The average HF researcher is probably unconcerned about this.

The measures compiled by Gawron (2000) offer a starting point to consider measures as these are actually used by researchers. The most common objective measure is accuracy (e.g., number and percent correct; Van Orden, Benoit, & Osga, 1996), but more often transformed into its converse, error. Error measures include absolute error (Mertens & Collins, 1986), number and percent errors (Hancock & Caird, 1993), error rate (Wierwille, Rahmi, & Casali, 1985), false alarm rate (Lanzetta, Dember, Warm, & Berch, 1987), and root mean square error (Eberts, 1987).

Many measures are assembled in the form of a battery consisting of a number of related tasks. For example, the Advisory Group for Aerospace Research and Development (AGARD; 1989) developed a series of standardized tests for research with environmental stressors (STRES). This battery consists of seven tests, including reaction time, mathematics, memory search, and tracking. A similar test battery is the Performance Evaluation Tests for Environmental Research (PETER), which is made up of 26 tests, including such disparate tests as air combat maneuvering, reaction time, and grammatical reasoning (Kennedy, 1985). Many such test batteries have been adapted to computer presentation. Obviously, there must first be a theory to arrange the combination of these tests into a single battery.

Such a test battery includes a variety of tests with a much more limited number of measures, such as error. Researchers rarely distinguish between measures, such as error, and the tests in which these measures are embedded. The difference between the two, tests and measures, may be of little practical significance, but the important point is that, although the number of measures is limited, the range of their application is much greater. As pointed out previously, any measure such as error can be applied to a wide variety of human capabilities (e.g., memory), human functions (e.g., detection), the comparison of different types of subjects, and different aircraft functions (Bortullusi & Vidulich, 1991). The emphasis in these tests is not so much on their nature (e.g., the complexity of their dimensions) as on their capability to differentiate between experimental groups (e.g., different environmental stressors, subjects with different amounts of experience, gender differences). Despite the fact that a theory organizes these tests into a single battery, their dimensional characteristics are often unknown, which does not, however, prevent researchers from applying them.

It has been pointed out that in developing a measurement situation in which the subject must respond, the researcher creates a momentary reality for the subject (i.e., it persists as long as the subject is being tested). When an actual system is involved, the system imposes its own reality on subjects.

With conceptual research, in which the measurement situation does not resemble a system, and with subjective measures and particularly if concepts are involved, situational information should be provided to the subject because there is no actual system to guide the subject's performance. That information and the temporary reality previously referred to are usually presented in instructions for subjects. Be careful, however, not to bias subjects by the context provided in instructions.

In effect, the subjects are asked to imagine a context in which they perform. For example, the subjects of a performance test assume a role (that of a test-taker), which may be vastly different from their ordinary life performance. The measurement specialist assumes that most people can assume their performance-measurement role without excessive difficulty; this is part of the specialist's implied assumption of the meaningfulness of the test outputs. In most cases, the assumption is valid, but some subjects may not be able to adapt to the special performance measurement (e.g., they are poor test takers); and their responses may not actually reflect their true performance capability (although those results truly reflect their present test performance). To reduce the impact of such deviations is one reason—not the only one, of course—for the use of variance statistics, such as the standard deviation.

Less information is available to the subject of a conceptual study, because system information (which is often voluminous) is not available. In consequence, subjects of conceptual studies may often have a more limited test context than do subjects of system studies. Experimenters assume that the experimental controls inherent in the treatment conditions in which subjects perform will compensate for subjects' lack of information. In other words, the treatment conditions with their constraints force the subject to perform in a particular manner; it is assumed that it does not matter, therefore, if the subjects do not know why or to what their performance relates. There is no hard evidence to indicate whether this last assumption is correct.

Measures provide diagnostic clues as well as evaluational data (i.e., how well a task is performed). For example, the type of error made (e.g., production errors) may suggest its cause. An increased frequency of error at a particular mission phase or in connection with a certain function or task may suggest causal factors, which are to be found in the nature of the function/task and how it is to be performed. Subjective measures may provide more of this causal information than objective ones, because the subjects may be more aware of their performance than in objective tests, and they may be more able to explain why a difficulty arose.

Measures may be multidimensional as well as unidimensional, and may even be derived from the development of an equation relating several molecular indices to a molar phenomenon (Richard & Parrish, 1984). It seems likely that the more elements that enter into an equation, the less will be known about the dimensional "purity" of the equation, because elements are intermingled. Each new measurement situation (e.g., measurement at different mission phases) may require the development of measures appropriate to that situation (Charlton, 1992). Both objective and subjective measures may be commingled in a multidimensional measurement situation.

Quantitative correctness scores can be assigned to activities, with the basis of the measures being a rating or scale based on observation. For example, Giffen and Rockwell (1984) evaluated human problem-solving efficiency based on a 5-point rating scale of how the subject performed (e.g., the subject asked for incorrect information or was on the right track). Scales have been developed to measure a wide range of situations. The basis of the scale measurement is the human's role in the function/task being evaluated. For example, in measuring display readability (Chiappetti, 1994), it is the operator's response to the display that determines its readability, and nothing else. The same is true of workload (Roscoe, 1984) and crew status (Pearson & Byars, 1956). The source of the performance is the human, which makes subjective measures and, in particular, scales and rating so appropriate.

Behavioral measures have been applied to team and organizational dimensions, as well as to individual performance (Nieva, Fleishman, & Rieck, 1985). If the system organization has multiple aspects, which are manifested overtly, it is possible to use these aspects as measurement criteria (Boyett & Conn, 1988; white-collar performance measures). If performance is determined by objective criteria, such as the number of personnel as against task requirements, it is possible to develop equations incorporating these dimensions that reflect the variability of team and organizational performance. The point is that behavioral measures need not be confined to the error dimension (as if there is only one correct way to operate a system). It is possible to use objective criteria of system performance (e.g., the organization's "bottom line") to evaluate performance. Indeed, such measures are used as much or more by nonmeasurement specialists to evaluate performance in real-world nontest situations.

Many of the measures recorded in Gawron's (2000) compilation were designed to measure workload, the concept being that an excessive workload is responsible for inadequate performance. Implicit in this concept is the more fundamental assumption that stimuli impose a demand on the human's resources, and those resources may prove inadequate to the demand.

Constructs such a workload, stress, fatigue, and complexity all contain multiple dimensions. Attempts to measure such constructs by abstracting a single dimension usually turn out to be unsatisfactory. A prime example is the definition of workload as the ratio between the time required to perform a task and the time available for the task: If the comparison results in a value above 1, then workload is excessive. Obviously, workload is more than time availability.

Researchers who recognize that complex constructs like workload are multidimensional develop multidimensional scales. One of the most widely tested of such scales is the NASA Task Load Index (TLX; Hart & Staveland, 1988), which consists of 6 bipolar scales (mental demand, physical demand, temporal demand, perceived performance, effort, and frustration level). Subjects rate each task on each of the 6 subscales, followed by 15 pair-wise comparisons of the 6 subscales. The number of times each scale is rated as contributing more than other scales to task workload determines the weight of the scale in the overall equation. An average overall workload is determined from the weighted combination of scale scores on the 6 dimensions.

Researchers would assume that there would be a relatively high correlation between the overall workload rating and different system phases, as in aircraft flight segments. For example, aircraft cruising should impose less workload than takeoff and landing. This is, of course, what has been found.

Nevertheless, the use of composite scales like TLX is a triumph of pragmatics over cognitive complexity. The elements of the equation may make sense as individual dimensions, but less sense when combined, because interactive effects are unknown. (Nevertheless, the further progress of HF measurement must inevitably be tied up with equations consisting of multiple elements, because multiplicity is the hallmark of real-world human performance.) Pragmatics are important, however, if the equations discriminate between conditions of interest. In such cases, usefulness determines validity. The use of the secondary task in workload measurement is a common example of research pragmatics; researchers do not really know how secondary tasks interact with primary tasks in real life.

Such divided attention problems are more common in less formalized, nonsystem situations. In any event, a rule of measurement in system studies is that any measure employed must not interfere with the basic system task. Nevertheless, the secondary task as a measurement device has gained considerable popularity within the research community.

There are continuing attempts to validate the more complex measures of constructs like workload, but not metrics. Metrics representing inherent performance dimensions, like error and time, need not be validated. For metrics, criterion of relevance takes the place of validity. Nevertheless, the error metric is meaningful only if the task being measured can be performed in only one way. If that assumption does not hold, then the error, as a deviation from correct procedures, is not meaningful.

Of course, error could be replaced with a measure defined by the alternative ways in which the average subject would perform the task. Thus, if a "reasonable" individual would not utilize a particular performance method, it would be incorrect. (How to define a "reasonable" individual presents another problem.)

On the other hand, where alternative strategies were possible in performing a task (e.g., in failure diagnosis), the relative success (involving a time factor, perhaps) in utilizing each strategy could supply weights to make each strategy a success or failure. This suggestion is made only to illustrate that complex cognitive strategies often do not lend themselves to unidimensional measures, and these strategies can be used as the basis for developing measures.

If researchers cannot operationally define a construct like stress or workload because of its complexity and the difficulty of defining how their individual dimensions interrelate, then the procedure seems to be to develop scales for the individual dimensions and then leave it to the subjects to supply the combinatorial equation by the weights they place on each dimension. Unless subjects are very articulate, researchers do not know what the subjects are defining for themselves in using the scale, how accurately subjects can recognize their internal status, and if subjects can match that internal status against the specified scale values. The subjects thus become an uncalibrated instrument to measure something that is not clearly evident.

Although reliance on an uncalibrated subject may not appear very rational, it occurs because of the implicit assumption that the subjects can recognize their status and can match that recognition to the specified scale values. The result is, as was indicated previously, pragmatically useful, and there is obviously some correlation (although how much is unclear) between the reported internal status and any "true" value of the construct.

Many tasks from which measures are derived may possess significant deficiencies. For example, Gawron et al. (1989) identified five major deficiencies in then current in-flight workload measures (cited in Gawron, 2000, p. 101). Many measures are overly simplistic, concentrating on only one dimension of a multidimensional task, or the task is overly complex, involving too many dimensions to deal with effectively in a single measure. Often researchers are not very clear about what they are either theoretically or actually measuring.

The point is that, in general, measures are far more complex conceptually (even if their use is apparently simple) than most professionals who use measures are aware. The influence of "customary measurement practices" makes it quite easy (and indeed desirable) to ignore this difficulty in performing measurements.

It would be too much to expect researchers to surrender the use of common (and not so common) measures in order to resolve conceptual difficulties. However, researchers could be asked to explain in the published paper why they selected a particular measure.

CRITERIA

Before selecting one or more measures, it is necessary to specify *criteria*. There are two major criteria in system performance: mission *success* and function *efficiency*. In conceptual studies that lack a system and mission context, the primary criterion is efficiency, measured by the subject's accuracy and time. In systems, the efficiency criterion helps to determine success, but cannot completely determine that success, because nonsystem factors such as weather or inability to secure needed information may reduce that efficiency. For example, in the Civil War, Lee's highly efficient Army of Northern Virginia became less efficient, because of poor weather and, at Gettysburg, it lacked information about the enemy's position.

In both system and conceptual studies, the system performance or the human subject's performance is related to what the system designer and researcher wanted the system or the subject to do. In the system, there is an actual requirement (e.g., the aircraft is designed to fly at 70,000 feet, and if the aircraft can fly only at a maximum of 60,000 feet, the success of the mission is imperiled). In the conceptual study, the requirement is implied; the task subjects are supposed to perform (e.g., operating a console in a specified manner) suggests by implication that success is represented by lack of error in the task sequence. In both the system and conceptual study instances, a goal is associated with a system or human function, and the goal is 100% efficiency. However, because the conceptual study does not explicitly require anything of its subjects, there is no standard and whatever performance is achieved is accepted.

In nonsystem-related performance (e.g., in subject performance of a laboratory visual detection task; finding all the circles in long sequences of *X*s), the common criterion is accuracy, which is performance relative to complete efficiency (100% detection of circles).

In systems there are three distinct but interrelated types of performance criteria: those describing overall system functioning, those describing how missions are performed, and those describing how personnel respond in performing system functions and missions. System criteria include such descriptors as reliability (e.g., How reliable is the system?); mission descriptive criteria include success or failure in system goal accomplishment, output quality, and reaction time (e.g., How well does the system perform its mission?). Personnel criteria include accuracy and speed (e.g., How proficient are operators in performing emergency routines?).

In nonsystem-related human performance, the criteria utilized are related only to the measurement task required of the subject. If the measurement task is synthetic (constructed only for measurement purposes), it is unimportant except as it describes the *function* involved in the measurement task. The researcher in the previous circle-detection task is interested in detection capability as an attribute rather than the specific task of finding circles in relation to *X*s. Task measurement in nonsystem-related situations is unimportant, but the function represented by the task is important.

Performance criteria may act as independent or dependent variables. As independent variables (e.g., the requirement to produce *N* units or to find all the circles in the *X*s) they impose on the system operator or experimental test subjects a requirement that serves as a forcing function for their performance. As dependent variables they describe the operator's performance in response to a requirement (i.e., the operator has or has not produced *N* units) and can be used to measure that performance. The latter is of most concern here.

In evaluative performance measurement (Do subjects perform as required?), the criterion requires some standard of that performance. In conceptual research, criteria are necessary to describe the effects of independent variables, but these do not imply or require standards.

General criteria like efficiency or number of units produced or transmitted are not sufficient. The criterion must be made more precise by becoming quantitative, which means applying the number or time metric to the criterion. Moreover, not all criteria are equally relevant and valuable for performance measurement. The physiological criterion of level of adrenalin in the blood of subjects performing a visual detection task may be related to target detection, but is not the most useful criterion because it is only indirectly related to task and system outputs. In a system context, the relation of a criterion to system mission success determines the relevance and utility of the criterion. In nonsystem performance measurement, there is no criterion external to the human performance, but it is possible to arrive at something similar by relating the performance criterion to the function (e.g., detection) involved in the specific task. Accuracy of detection in a laboratory test can be considered equivalent to system success, but only on a personal level.

ASSUMPTIONS UNDERLYING MEASURES

To understand the use of measures, it is necessary to peer beneath their surface and examine the assumptions that underlie them. Some of the following assumptions apply as much to measurement in general as to measures specifically. That is because there is an inextricable interrelation between the measurement process as a whole and the measures used to describe performance.

Performance Variability

The notion of performance measurement assumes that there is a probability of some deviation between actual (measured) and desired (designed) performance, as the latter is described in manuals. The procedure described in the manual assumes and requires 100% efficient performance, but measurement of operators performing that procedure indicates an average error rate (deviation) of .003. Only performance measures will indicate this deviation. If no such deviation was anticipated, then it would be unnecessary to measure performance at all. An example of the concept of deviation is error, which defines itself specifically as a deviation.

A related assumption is that a certain amount of variability is inherent in all human performance. The standard deviation curve is a common example. Only after performance variability exceeds the "normal" (whatever the majority manifests) can causality be ascribed to the factors supposedly determining that performance. For example, three standard deviations beyond the mean is commonly accepted as very exceptional performance.

This assumption is the basis for the statistical significance levels used in experimentation to test differences between treatment conditions. The norm (whatever norm is accepted) becomes the criterion against which all performances are compared. If individual differences from the norm become excessive, then these must be explained away.

Error

The common error measure assumes that there is only one correct way of performing a task. Any deviation from that right way represents less than 100% efficiency.

As has been pointed out, error is a binary concept; either an error is or it is not made. This is true, however, only if the task being performed is highly proceduralized so that error can be categorized in terms of a specific deviation from the procedure: errors of omission, errors of commission, or performance out of sequence.

If the common concept of error is meaningful for highly proceduralized tasks, what can be said about error when the task is nonprocedural, when there are alternative ways of achieving the task goal, and when the error is insignificant in the accomplishment of the task goal? In a procedural task, it is almost inevitable that an error of any type will result in task failure. In other kinds of tasks (involving contingencies, which means they cannot be proceduralized), the error may not prevent the operators from accomplishing their goal. In system tasks, the goal of performance is not merely to perform the task; the task is performed in order to implement a higher order purpose. Under these circumstances, the measure must be related to that higher order purpose. In conceptual studies, the performance may appear to be its own justification, but this is only superficial; the measure must be related to the *function* of the task and/or to the *variables* that the measurement studies. Only then does the measure make sense.

Consider that only an error that prevents goal accomplishment is of consequence; if it does not, then the error is merely descriptive and not causal. The error by itself means nothing if its consequences are not considered.

Some errors can be canceled out if the operator has time to recognize the error and to perform again the action that led to the error, this time correctly.

In actuality, then, the concept of error is not monolithic: Errors depend on the nature of the task, the consequences of the error, error recognizability, and error retrievability.

Error depends on knowing what is correct, what the standard is against which any individual action can be compared. If someone does not know what is correct, then can there be an error? If the purpose of task performance is to achieve a designated goal, then it would make more sense to designate an error as an action leading to nonaccomplishment of the goal.

Thus, any error would have a probability of goal nonaccomplishment attached to that error. It would still be possible to count errors and to use them to differentiate treatment conditions, but the meaning of the error would be more complex than simply as a procedural deviation.

Error as a probability of goal nonaccomplishment is not the same thing as the more common error rate, which is the ratio between opportunities to perform an action and the number of times that action resulted in error. Error rate includes all errors, whatever their effect on goal accomplishment. In complex nonproceduralized tasks like investing in the stock market, the underlying assumption is that one strategy is more likely to result in profit than another strategy, but that no strategy has a 1.0 probability of success. Thus, errors might be defined as actions that have a greater than 0.5 probability of failure. If in a nonproceduralized task there is no absolute probability of success, it might be more meaningful to talk of actions having a success probability, rather than of error.

The Goal

Another concept underlying performance measures is that the task or system goal is the driving force behind individual and system performance. Accomplishment or nonaccomplishment of the goal can therefore be used as the overall performance measure to which all other performances (and the measures used to describe these) are subordinate and contributory.

In any system with multiple levels, the actions involved in the various levels determine the specifics of goal accomplishment. Each level has its own goal that must be related to the overall goal of system success. Goals are physical and behavioral; the overall system goal is physical; subsystem and task goals are of both types. The behavioral goals are linked to tasks and functions performed by system personnel. In conceptual studies, there are fewer goals and these are associated only with the individual task and function. If subordinate system goals do not relate to the overall goal, then the system is poorly designed. This is more likely to occur in commercial systems that have a less formal organization than, for example, those of a military or power generating nature. The fact that human–machine systems have both physical and behavioral goals means they must be interrelated.

The task always involves an equipment manipulation; this makes the interrelation at the task level relatively easy, because the behavior subsumes the equipment function. As soon as multiple tasks are involved in actions, the physical/behavioral relation becomes confusing; the individual task–multiple task relations become obscure, because individual tasks may not have the same degree of relation to the higher order physical mechanisms.

The combinatorial processes suggested by human reliability theory are an effort to make quantitative sense of these relations (see Kirwan, 1994).

Molecular–Molar Relations

In studying molecular behaviors in conceptual research (e.g., individual functions like detection or discrimination, as in the task of finding circles in a sequence of *X*s), the assumption is made that such molecular functions combine with other functions in a task context. These, then, influence the more molar task performance and the latter affects system operations. If this assumption were unwarranted, there would be no point to performing conceptual studies describing individual functions. To consider functions independent of a task is pointless, because only tasks can influence the operator and the system's performance. A decrement in human function, like detection, may occur, but these have no effect on the individual or the system until they affect the task. Only the task can influence the individual's or system's performance, because the individual's goal animates the individual and the system, and the goal applies only secondarily to the function.

Because there is a difference between molecular and molar human performance, it is logically necessary to try to show the relation between the two in quantitative terms. What, for example, is the relation between throwing a wrong control on a radar set and the performance of the radar subsystem in the context of total ship operations? It may not be easy to trace the relation because the molecular function, such as detection or recognition, is subsumed by the more molar task (e.g., radar tracking of a target). The notion of conceptual research of molecular functions, like detection, implies that whatever is learned from such studies has some application to more molar behaviors of the task/system type. Such an assumption should be tested by finding a relation between the results of the experiment and those of the system test. The effort should be made (perhaps by means of a correlation between molecular and molar aspects of system performance), because if there is no discernible relation, a great deal of effort has been wasted studying molecular behaviors in conceptual research.

Measures differ in molarity. The most molar objective measure is error, because it is a consequence of all the more molecular behavioral functions, like detection and perception, with which it is linked. Time is a more molecular measure because it is unidimensional and only occasionally summative. A composite measure like workload is the most molar, because it is composed of several submeasures.

The importance of the assumption that molecular functions significantly influence overall system personnel performance is that it provides a bridge between conceptual research (mostly molecular) and system (mostly molar) performance. A conceptual study can investigate these more molecular behaviors directly. This cannot be done in a system study in which molecular behaviors are bundled into more molar tasks, hence the utility of the conceptual study.

Other assumptions have been discussed in earlier chapters. One assumption is that there must be no interference with system operations while these are being measured, and where any interference might result in arbitrary stoppage or delay of those operations by external—meaning researcher—actions. Because of this assumption, objective behavioral measures (which do not depend on subject interrogation) can be used in both system as well as conceptual studies, but self-report measures (which rely on some form of subject interrogation and, hence, influence on the subject) cannot be used in system studies until after the system has completed its normal mission.

Use of the time measure assumes that time is a critical dimension of the particular human performance being measured. Otherwise, there is no point to taking time measures.

A pragmatic assumption, because its purpose is to aid measurement, is that not all performance needs to be measured. Most human behaviors, although assumed to be necessary to perform a task or mission, do not contribute equally to system success. Application of a measure to a particular performance means that the researcher assumes that a certain performance is more important than some other performance to which the measure is not applied.

It is assumed that human performance in a system context contributes a certain (presumably quantifiable) increment to accomplishment of the system goal. If this assumption were not true, there would be no point in measuring the performance of system personnel. Unfortunately, researchers can prove this only in a negative sense (i.e., as when an error or other behavioral deviation results in mission failure or catastrophe). Nothing, except that its design was adequate, can be learned when the system works well. How much is contributed by personnel depends on the nature of the system and the human role in that system. Where the human role is preeminent, as in an aircraft, the human contribution will be proportionately greater than where the human role is minor.

System performance is defined in terms of mission success (goal accomplishment). Therefore, any human performance measure that cannot be related to the system success measure is irrelevant to understanding of system performance. This does not, of course, apply to conceptual studies. In these studies, no system is presented, but rather function or tasks that could be part of personnel performance in a system. Success in a conceptual study is an individual accomplishment, not a system goal accomplishment. Because the system is not ordinarily involved in conceptual research, the performance that is studied in the latter describes a capability to perform a human function that may be required by a specific set of system tasks. The utility of the conceptual study is that it should and sometimes does produce capability data (e.g., minimal thresholds) that can be applied to practical purposes.

A related assumption discussed in previous chapters is that humans, considered from the HF perspective, do not function outside of some human–machine systems, however remote from them physically. Consequently, if nonsystem (conceptual) studies are performed, these can only be justified by whatever information they provide for system design and operation.

Behavioral measures taken outside of a system/equipment context cannot be used to describe system personnel performance until the measures are transformed into task terms. For example, visual acuity is measured using a Snellen chart. The chart measures visual acuity, not actual visual performance as it occurs during task performance. Hence the Snellen measure and its data cannot be used in a system context, unless its measure is transformed into one that is system relevant, perhaps by relating it to a function like the perceptual discrimination involved in a task.

QUESTIONS ASKED IN MEASURE SELECTION

The answers to a number of questions should help direct the researcher's attention to the most effective measure:

1. What is the nature of the function and task being measured? What are the implications of this for the measure? What is the role of the human in this function or task?

In a system study the centrality of the operator role determines how much attention will be paid to study the subject in the measurement. For example, the performance of flight attendants on an aircraft might not even be measured, because their activities are not central to the operation of the aircraft. In all studies, certain activities are more focal than others and the former will therefore receive more measurement attention.

The more functions operators perform during the mission (those relevant to mission success), the more measures of their performance have to be considered, even if in the end only one measure is selected.

2. What constraints reduce the number of possible measures?

The most obvious constraint in system studies is that ongoing system operations forbid interference with the personnel involved. Hence a measure that would produce this interference would not be used. Physiological measures that require obtrusive instrumentation might not be selected, unless there was no other way of gaining the desired data. In a conceptual study, in which no system is involved, the experimenter has more freedom to adjust methods of subject response. For example, in situational awareness studies, the presentation of stimuli can be interrupted to require subjects to answer the researcher's questions about those stimuli.

A primary constraint is the cost associated with the use of a particular measure. Every measure involves a cost represented by the way in which the information is collected. For example, errors may be recorded automatically or by observation; a performance can be observed manually or recorded automatically by camcorder. Where instrumentation is required, there may be financial costs or intrusion in subject functioning; where data collection is labor intensive, there will be labor costs. The researcher's natural inclination will be to select those measures whose data are easiest to collect and that place the least burden on research resources.

3. What effectiveness criteria should be applied in measure selection?

Some measures will be more effective than others. The question concerns how to determine the most effective measure. The primary criterion is relevance; all other criteria listed previously can be accepted or rejected in part as the measurement situation requires (e.g., a measure that is quite costly can be accepted if no other alternative to that measure exists), but only the relevance criterion is absolute. An irrelevant measure is unacceptable under all circumstances. One way of determining if a measure is relevant is to ask how data reflecting that measure would explain how anticipated performance effects occurred. For example, if researchers were contemplating use of time as a measure, they should ask how the time measure would explain expected performance. If time were not a critical factor affecting performance, then researchers would reject time as a measure. Other measures (e.g., consumption of resources; see Table 7.1) would also be rejected in most conceptual studies on the same grounds.

MEASURES AT VARYING SYSTEM LEVELS

One aspect of measurement and measures not yet considered in detail is their relation to system levels. It has been pointed out that the operators in their work almost always function in the context of a system; therefore, the relation between measurement and system levels becomes more than theoretically important. Researchers wish to know how performance at the workstation or subsystem level influences overall system success, because to influence the latter it may be necessary to modify the former.

Any system of any size consists of various levels, each of which contains a number of subsystems. The lowest level is that of the individual performing a task, in most cases at a workstation. If the operator is part of a team (which is often the case), then the individual performance must also be related to that of the other team members.

In measuring performance, these individual team functions must then be related to one or more subsystems at a higher level. Eventually, performance at subordinate subsystem levels must be related to the terminal, command, or executive level, the output of which represents the overall system's accomplishment, or its "bottom line."

This is a deliberately simple description of a system. At any one system level, there may well be a number of subsystems functioning in parallel, each of which is subordinate to some higher level subsystem; and so it goes. For example, the highest level in a destroyer is the command/control subsystem (the "bridge"); inputs to it are received from subordinate communications, radar, sonar, weapons, and so on. These subsystems have one or more workstations, and at the lowest level, the individual sailor. Military systems are, by their nature, very hierarchical, but most civilian systems are similar, except for the command formalities.

Theoretically, at least, all functions within a system are necessary, but some functions are more important than others that are or can be delayed or performed only partially without highly negative effects to the system as a whole. A warship, for example, cannot function without a galley (cookhouse), but weaponry is more important in terms of the ship's overall goal, which is to destroy an enemy.

The point is that what is measured is as important (or even more important) as how it is measured. Given that everything in a system cannot be measured (except when the system is very small), a choice must be made among each of the tasks and levels within the system.

The end product (goal accomplishment) of the entire system depends on both equipment and human elements. The measurement of system output consists, therefore, of both physical and behavioral inputs. Because the measurement of system effectiveness combines these disparate inputs, it is necessary to show how they are related. This end product (system effectiveness) is described in a single measure and does not differentiate between the contributions of equipment and human elements, as may be seen in the following sample measures of effectiveness taken from a handbook describing operations research measures (Rau, 1974, p. 15):

(1) cumulative distribution of maximum detection range based on the fraction of runs on which detection was made by a given range;

(2) single scan probability of detection as a function of range for given target speed, target type, antenna type and tilt angle. This is measured by:

$$\frac{\text{number of detections (blips)}}{\text{number of scans.}}$$

Obviously, the human, although contributing to the system index, is completely subsumed by the physical measure.

The question of the human contribution to system operation is a very practical one. The human is necessary to the performance of the human–machine system, but in the mind of the system development manager the human contribution may be viewed as much less important than that of the equipment. If the HF design specialist is to assist effectively in designing systems that make maximum use of the human, then the importance of how the human influences system performance must be demonstrated quantitatively.

One of the significant research tasks for HF is, therefore, to determine how to measure the human contribution (taken as a whole, which means including all relevant personnel) to the ultimate end system product. Determination of this contribution may involve tracing the effects of error or failure to perform effectively in the individual or the single workstation from one system level to another and up to the terminal system level. It is assumed that these effects proceed not only vertically (i.e., from one subordinate level up to the command level), but also horizontally (in terms of interrelations between subsystems on the same level), and downward (commands from higher to lower levels). The example is in terms of military systems, but civilian systems function in much the same way.

Each system level presents a slightly different reality picture for the personnel at that level. The sonar person at the sonar station is absorbed in the elemental picture of visual displayed sound impulses hitting a target; the commander or executive officer on the bridge is concerned with a more complex reality (How should the destroyer maneuver in relation to a target at such and such position?). Each picture is real, but contains only a slice of the total measurement situation.

What changes importantly at each subsystem level is not the number of personnel involved, or the complexity of the subsystem equipment, although these may vary. Rather, it is the significance of the human performance at any one level for the system as a whole. For example, if the ship commander orders a change in direction for an attack on a submarine, the mechanical agency (helm and engine movement) remains the same; the behavioral (performance) functions of the helmsperson are the same. What changes are the effects of ship movement on the entire system in relation to the system goal. The ship commander made a decision (performed a behavioral function) which, transformed into a command, influenced first the helmsperson, then the ship's engines; but the effects of these, although similar in form to any other command decision, now change the functioning of other system elements in terms of the overall goal.

It seems a reasonable hypothesis that one way of tracing the effects of a behavioral action like a command decision on the system is to measure its effects in relation to the accomplishment of the system goal, which is the only fixed element in the process.

But, what if such a relation cannot be traced? It is not easy to decompose the system level measure (e.g., probability of killing an enemy, accomplishing a corporate merger) into its constituent elements. Researchers can, however, try at the least to show a parallel relation between variations in system success and variations in human performance. It is, in any event, necessary to try, because if they cannot show a relation between human performance and system success, those who design and control systems will feel that human performance, although necessary to system functioning, is not really determinative of that functioning. This will cause them, then, to pay less attention to the human factor in system design and operation than they would otherwise.

Tasks and measures change at different system levels. Performance of a function at one level will change into performance of a different function at a different level. For example, performance at the operator level is phrased in terms of error, range at which the target is detected, and so on. The commanders are, however, not concerned with individual operator error or the operator's detection range; they are concerned with the estimated range to the target (which combines the operator's precision with the machine's precision). The command perspective, regardless of system level, is on system performance (not operator, not machine) except when either of these two last fails catastrophically.

It goes without saying that the measures applied should be appropriate for the system level at which performance is to be measured. Rau (1974, pp. 6–7) suggested that as the measure of effectiveness (MOE) hierarchy evolves from the top level to the lower levels:

> The nature or form of the MOE changes. At the lower levels, the MOEs become less "effectiveness oriented" and more "performance [goal accomplishment] oriented." For example, median detection range, circular error probable, mean miss distance, etc. are typical effectiveness oriented MOEs, whereas expected number of target kills per sortie, [and] probability of target detection, . . . are typical performance oriented MOEs.

Some metrics, like time and error, are appropriate at all but the terminal system level, but other measures at a lower system level (e.g., probability of operator detection of a radar target) must be transformed at a higher level into another measure (e.g., probability of an enemy "out there" or probability of an enemy kill). This change in metric increases the difficulty of disentangling the human contribution from the overall system measure.

System level relations may also be understood in terms of dependency. For an effect of an input from one level to manifest itself at a higher level, the dependency relation of the first to the second must be ascertained. If there is no effect of one on the other, there is no dependency (and vice versa). For example, the choice of a menu for the evening meal has no effect on ship maneuvering, so no dependency exists. However, most performances at different system levels will manifest some dependency. It then becomes necessary to ascertain the degree of dependency.

Because of these system interrelations, it might be said that a measure is effective to the extent that it reveals changes in performance (e.g., number/type of decisions) over time and over various system levels. The absence of change may be an important indicator (like Sherlock Holmes' dog that did not bark in the night), but a flat curve is less interesting and usually less informative than one that shows variations.

How would one go about attempting to study the human contribution to system performance? It would be highly desirable to have a theory, however inadequate, that would attempt to explore this relation. No such theory exists, because the concentration of HF intellectual effort is not on the human relation within the system, but on the human response to system elements.

The development of such a theory requires the gathering of descriptive data on how systems and humans perform. This can be done by measurement of actual systems functioning in the operational environment, but this presents problems. Even a descriptive approach to system functioning requires the development of questions to be asked. The ultimate question is, of course, the one that initiated this discussion: How does personnel performance affect system operations? To answer this question, more discrete, more elementary, molecular questions such as the following need to be answered:

1. What is the nature of the data inputs to system personnel, and how do these affect system functioning? What reports are asked of the operator and what happens to those reports? What do system personnel do with these reports?
2. What is the effect of human performance deviations (e.g., errors, time delays) on the performance of other system elements?
3. Can investigators graphically trace the effects of communicated information as that information proceeds up and down system levels? Information is the medium by means of which effects are transmitted through the system.

Undoubtedly, many other behavioral questions will arise following observations of actual system performance.

The primary difficulty in answering these questions is that there must be an actual or simulated system to observe and measure. The opportunity of observing actual systems is limited by the reluctance of system managers to permit researchers to observe system operations. That difficulty is often insuperable in civilian systems, but can sometimes be overcome in military systems where governmental agency influence can be exerted.

If observations are possible, they will probably center on communicated information, communication channels, information effects, and resulting system performance changes.

Because the opportunity of studying actual systems in operation is slight, the need for human performance models and system simulations becomes apparent. The data acquired by descriptive observation and measurement of operational systems will then become inputs to the models. Of course, this raises a Catch-22 paradox: If the opportunity to observe actual systems is lacking, how do researchers go about getting the input data needed to exercise the models? Obviously, compromises involving a good deal of extrapolation of available data will be necessary.

If input data are available, then it will be possible to exercise the models and manipulate input data variables. Effects on ultimate system performance (in terms of accomplishment and nonaccomplishment of the system goal) can then be observed.

This is a very simplistic description of a very involved measurement process, certainly not one that could be accomplished by a single investigator in a single study. The process is possible, of course, but only if sufficient enthusiasm for the work can be generated among funding agencies. The likelihood of this occurring is not great, unless HF professionals as a whole can convince these agencies (after they have convinced themselves) that HF is a system science and not merely the study of psychological reactions to technological stimuli.

The study of human–machine systems is not easily reduced to the constraints of ordinary experimentation. It is much simpler to pursue traditional conceptual research studies involving relatively molecular aspects of the system (e.g., icons, menus) and the responses of individuals to these aspects. The implicit assumption behind this strategy is that research on molecular technology aspects will eventually add up to knowledge of the total system. This is very probably a weak assumption. The present research on individual system aspects is necessary and may provide the guidance required by HF design specialists, but it will not summate to an advanced understanding of systems as a whole.

Chapter 8

Research Users

As soon as its results are written down, research assumes the additional function of communication. Communication has two interactive elements: one or more who communicate, and one or more who receive the communication and do something with it.

The two are equally important: The communication must be a stimulus to further action, unless the communication is irrelevant to any ongoing or further function, or is ignored. Indeed, unless the communicator receives feedback from the receiver as to the action resulting from the communication, the communication is essentially impotent. This implies a difference between communication and the influence of that communication; this is reflected in the difference between communication without a change in the recipient's performance, and communication followed by such a change. This is true, of course, only in the short term; "knowledge" can be thought of as a form of communication, or at least mediated by communication, that functions over time, even when the original communicator is long gone.

RESEARCH UTILITY

In this context, the assumption is that research has its uses. If so, a discussion can talk of a research user (the receiver of the communication) and examine the uses to which that research is put. The research user is anyone who is affected, however slightly, by the communicated research output. Operationally, this user can be defined as someone who reads the communication (because most communications of a technical nature are presented in written form, although they may be preceded by a verbal presentation in a conference).

The HF professional does not ordinarily think of research as an object for which there is a user or consumer. The intent of conceptual (basic, nonapplied) research is supposed to be to add to the store of scientific knowledge. Knowledge is an abstraction that does not appear on the surface to require interaction with a user. System-oriented (applied) research has in contrast specific problems and uses to which the research will be applied in anticipation of the actual conduct of the research. However, the uses and users of system-related research are much more private, because this research is not usually published in the general literature.

Communication, publication, and the uses to which the research is put are essential elements of measurement. If all HF professionals who read research studies were suddenly unable to understand these studies (a sort of research tower of Babel), imagine the impact on further research!

Research outputs that are not communicated in some way (usually by publication) can have no general effect on others, although they may have very useful private effects, as in improving system design and operation. Measurement not performed for general dissemination affects only the researcher who performed the measurement, and those who paid the researcher.

Is it possible that if the user cannot make meaningful use of the research communication, then the research has failed and the researcher's efforts are nullified? The concept of research use and user assumes that the research has a goal, which includes communication, and so, by generalization, the research output also has a goal. If so, the research and its use can fail if that goal—communication—is not accomplished.

If it is assumed that research is ultimately a communication, then the ability of the research to satisfy user needs determines, at least in part, research effectiveness. This makes research a product somewhat like any other product, which must satisfy its users if it is to accomplish its function. If research literature were a commercial product like a tube of shaving cream, those who created that literature would have to take into account the readers' interest in the literature, their need for information of a certain type, the effort involved in reading and analyzing the literature, and so on.

But, do researchers think of the reports they write as communications, as involving a reader whose characteristics and reactions to the report are important? The partial answer is, yes. At the least, researchers must make their message clear. Moreover, the researcher must be persuasive in order for readers to accept their interpretation of those facts. Among other things, this means no excessive speculation, and no claims that cannot be buttressed by evidence.

Researchers must be persuasive because they began the study with hypotheses that, because they were only hypotheses, had to be tested to confirm their verity (hence use of the experiment as a testing methodology). Even after the study is completed, the statistics analyzing the data merely demonstrate that the results were unlikely to have occurred by chance. In consequence, even with data supported at the .001 level of significance, there is a certain amount of uncertainty about the truth of the findings, particularly of the conclusions inferred from the data. The possibility always exists that the report is in some error. All study results are only probabilistic, until, in the fullness of time, they are supported by additional data; hence the need for the researcher to be persuasive against possible doubts.

Readers are not a fixed quantity either. They have biases, stemming from training and experience, which the researcher should take into consideration. However, so little is known about these biases that it could be difficult for the researcher to consider them; this suggests that there should be some systematic study of readers.

The researcher is not concerned about all readers. The researcher is not, for example, concerned about readers who are laypeople in the specialty area discussed. The opinion of general readers is worth little, because they are not specialists in the area under discussion. Only the specialist reader is important, because the specialist may cite the research in a later report, and the more citations the researcher can accumulate, the more prestigious the paper and its author. Moreover, peer-reviewers are likely to be specialists also.

The acceptance of a paper by peer-reviewers and then by the reader in part depends on whether it possesses a number of characteristics. The most important of these is factual: those questions the study has actually answered. The following characteristics are less substantial, more aesthetic, responding primarily to reader predispositions. None of these latter characteristics fully determine a paper's acceptance, but together they add to the general impression the paper makes and to the likelihood of its publication. These nonfactual characteristics include:

1. *Length.* A longer paper (within bounds, of course) is more convincing (respectable?) than a very short one.

2. *Statistics.* A paper with statistics (the more complex, the better) is more admired than one without statistics or with only minimal ones (e.g., frequencies).

3. *Diagrams.* Diagrams in a paper suggest a greater degree of logic and sophistication in the paper's argument than may be warranted.

4. *Citations.* The more citations in the paper, the more the researcher is assumed to have "consulted the authorities," and the more authoritative the paper appears.

5. *Theory.* Mention of or inclusion of a theory tends to a more positive reading by the reader, because theory making is supposed to be some evidence of intellectual sophistication.

6. *Lists.* The formulation of items in a list (no matter what those items are) tends to make the paper seem more logical and factual, even when no facts are presented.

7. *Theme.* Certain topics are more admired at certain times than are other themes. Those topics that are new and "hot" (like "situation awareness," Endsley, 1995) are more likely to be viewed favorably than topics that have been well mined in the past. However, "hot" topics cool off quite rapidly.

The aforementioned characteristics are only hypotheses, based on experience and observation (with a dash of cynicism added).

The preceding assumes that measurement/research has utility, which raises the vexing question of what this utility is. Some professionals will not agree that research has utility. If the ultimate purpose of the research is, in their minds, to generate knowledge (whatever knowledge is), and if knowledge is outside the domain of the quotidian and the normal, then there is not much point in talking about research utility. On the other hand, knowledge may have utility, except that, as in connection with communication, the utility may not be immediate. Probably most conceptual researchers assume that someone will come along eventually to retrieve the knowledge they provide and make use of it.

If knowledge is above "the madding crowd" and functions in a superempyrean domain, an investigator is permitted to ignore contraindicating factors. Indifference to research users is a product (in part) of graduate university experience, where almost all attention is given to doing the research. The university as the source of future knowledge makers is highly elitist, and so the user, as a dependent, is almost never mentioned. One aspect of research use, emphasized in the university, is, however, the research report's clarity of expression. This is important both in the university and thereafter, because what cannot be understood cannot be accepted. Researchers are willy-nilly concerned with research uses when they submit a paper for publication; editorial reviewers will not accept the paper if they do not understand it. However, even with editorial review of the new paper, the review does not take into consideration, except remotely, who the ultimate research users might be, or what they might wish to do with the paper.

Research/measurement is always performed in some context, which should include the goal of the measurement, the meaning and value of the research product, and the user of the research output.

Another sort of research use may be found in the funding agency. The reality is that funding choices determine almost entirely which research is performed and how. The funding agency may, therefore, be considered a user of the first instance. There is also the research community, defined as the entire cohort of professionals concerned with the researcher's specialty area. If, for example, other professionals decline to accept a researcher's findings, then the research might just as well not have been performed.

To require researchers to consider the value of their immediate research may appear to burden the researcher unduly. To place value on research may strike some as revolutionary. Knowledge has been supposed to have an inherent value independent of anything anyone may think or say about it. For example, to be ignorant of HF, as most laypeople are, does not reduce the value of HF.

To bring the user into the equation is to add an element of evaluation in relation to research. To be a user of something implies that the something is minimally acceptable. Acceptability implies that the product satisfies some criterion of acceptability, and a criterion requires evaluation.

If users are important to research, if the effects of research are important, then failure to utilize the research means that the research has failed, whatever other evaluation criteria are imposed.

Who Is the Research User?

The research user is anyone who funds research, reads the research report, and endeavors to apply research findings. The term *reader* is used to describe users in general, because the research must be read before it can be used. Being a research user, however, involves many more functions in addition to reading.

The class of readers can be further subdivided into categories, as follows: (a) *other researchers* in the same specialty area as that for which the study was performed; (b) the *casual reader*, interested only in browsing anything "new" or interesting in the discipline; (c) the *problem solver* (a reader who seeks help to solve a problem for which the study may supply information), a category including HF design specialists and funding agency personnel; (d) the *HF student*, who is required to read certain papers to pass examinations.

The categories are not mutually exclusive; one type of reader may become another, even concurrently. However, most HF readers of most papers at most times fall into the casual category, because they are not specialists, except in one or two areas other than the paper they are reading. Casual readers may also become problem solvers, if they must later solve a problem requiring information that the paper may possess. Students often read casually, and may, in their required research, also become problem solvers. The specialist reader is also a casual one.

A further distinction must be made between *present* users (at publication time *T*) and those of a *future* time frame (publication time *T* + 5, 10, 50 . . .

years; although readers may be both during their lifetimes). When a paper is read beyond the immediate future and particularly if it is utilized (at the very least, cited) in some way, the paper may become part of the store of HF knowledge. The concepts of the store of knowledge, and of research adding to it, can be operationally defined: Any published research that is read more than casually after its publication time frame, and is utilized beyond that period, is part of the knowledge store. Not every paper becomes part of that knowledge; indeed, the category may contain relatively few exhibits.

Beyond its existence over time, what makes a research report likely to become part of the knowledge store is its *utilization.* That utilization is of two types. In the immediate time frame (because the development of technology is likely to overtake research applications quickly), it is application to design of new technology. In a somewhat more extended time frame, it is use of the research as a stepping stone to further research. The latter use will also eventually be overtaken by the march of research; new research always replaces old research. In HF, the first application is often ignored; the second application can only be anticipated.

Obviously, there are papers that do not satisfy the knowledge criterion. It would be highly desirable to identify these before their publication; otherwise they function only as noise in the knowledge system. In all eras, studies have been ignored or found over time to have led nowhere, or have led the discipline in a less than optimal direction. Much earlier research may have been considered initially as part of a knowledge base, but have lost that status as new concepts and information were developed. An example is the "science" of phrenology or the chemistry of phlogiston.

If research application is necessary for research findings to become knowledge, then the researcher has the responsibility of showing, at least to some extent, how that application can be accomplished. For conceptual (nonapplied) research, the researcher should, as a minimum, indicate the questions the present research has answered (or failed to answer), and to what extent; those questions that remain to be answered; and how further research can make use of the present research.

The reader may ask why so much consideration should be given to the concept of knowledge, because the latter is largely a philosophical issue. The pragmatic answer is that "adding to the store of knowledge" is used as the ultimate rationale for the performance of much, if not most, HF conceptual research (a rationale of last resort, as it were). If the concept is not important to the professional, neither is the research it rationalizes.

The focus of the reader's interest in the literature may be general or specific, diffuse or highly focused. Readers who are interested in whatever is new have a very different form of concentration on the material, analogous to that of a newspaper or magazine reader. Those attempting to find an answer to a specific question will be more focused.

The type of reader interest in the literature determines the value of that literature to the reader. Manifestly, if readers are merely browsing, the literature can be almost anything; such readers make few utility demands on the material, which can then be evaluated very grossly. When readers demand specific answers, the literature will be held to a much higher standard.

Questions to be Asked of the Literature

Certain questions are implicit in any examination of the HF literature. First, why was the research performed? In general, there is a lack of knowledge about a particular topic, a lack that must be satisfied. Or, perhaps the available data are obscure, there are contradictions, or the researcher wishes to integrate the available information into a single comprehensive framework. Underlying all of these, but unstated, is the assumption that the research will make a contribution to the state of HF knowledge.

This applies to conceptual research. The questions are more straightforward for system-oriented research: Does the system perform as designed? How well do personnel perform? Of two or more system configurations, which is best on a human performance basis? What new functions or characteristics does a new system display?

In addition, how do professionals make use of the literature? For example, what is the frequency of their reading of the material, and the availability of their access to documents? How do they search for answers to behavioral questions? How do they feel about the literature? What characteristics of the literature most appeal to them?

Finally, how effective is the literature in accomplishing its purposes? This is a question that demands the prior development of effectiveness criteria and methods of measuring that effectiveness. This is the crux of any literature evaluation, but the problem is too extensive to deal with in this chapter.

Methods of Presenting Research Outputs

Research literature is not monolithic. It appears in various guises. First, it may begin as verbal presentations in conferences or symposia. The audience for this direct interaction of research output and user is somewhat limited, because not everyone has the opportunity and resources to attend such meetings. In addition, the interaction of listener and "presenter" via questions, answers, and comments, is very abbreviated.

Also, the verbal presentation is often printed and distributed as conference proceedings. This document reaches a wider audience among those who attended the conference but may not have had the opportunity to listen to the verbal presentation. In addition, copies of the proceedings may be sold to libraries.

Moreover, papers, whether or not they have been presented prior to publication, may be submitted to and sometimes published in peer-reviewed journals like *Human Factors, Ergonomics,* the *International Journal of Man–Machine Studies,* and so on. However, the rejection rate for submissions to a journal like *Human Factors* runs as high as 80%. Bound volumes of those papers that are published can be found on library shelves, and are available to anyone who wishes to read them.

The individual papers presented at meetings, together with published government reports, may be abstracted and presented as summaries in computer files, which are often searched by potential users in order to find material relative to their interests.

Finally, the available literature, both past and present, may be reviewed by specialists in various HF fields and commented on or at least cited in special review issues in journals or in handbooks (compilations) as reference material. If users wish to learn about signal detection theory or methods of investigating industrial ergonomics problems, they almost certainly will make some use of available handbooks like Salvendy (1987, 1997), Woodson, Tillman, and Tillman (1992), or Boff and Lincoln (1988).

Factors Affecting Literature Use

The first of these is physical availability of the literature (e.g., a library) and/or its *accessibility*; the latter interacts with type of user interest. If the user is motivated only by simple curiosity, then there is little problem of access. Readers simply browse as their interest and the availability of material permits. If individuals subscribe to a particular journal, they read the table of contents and select interesting material to read. They are not looking for any particular item of information; whatever comes along fortuitously is welcome.

If, however, individuals are attempting to solve a problem or answer a specific question, access is not immediate, because they have to extract a particular item from the masses of other material that are irrelevant. Simply to read a study about Topic X is easier than to find the paper(s) containing the data to solve a design problem related to Topic X. Presumably a computer search will isolate all that is known to have been written about Topic X (intermingled with a great deal of other material only tangentially related to Topic X). The information extraction process for Topic X, although tedious, can eventually be accomplished, if individuals have sufficient patience.

The nature of a problem for which the literature may (or may not) provide an answer focuses the reader's attention tremendously. Not every item of information about Topic X will be acceptable to the reader; information about X must now be very specific, otherwise it will not solve the problem. Users who examine the literature to solve a problem must define their questions sharply; otherwise the desired information, even if available, may not be accessed.

A very practical matter is the availability of libraries to the potential user. When budgets are tight, subscriptions to less well-known journals may be cut off. The professional who is not part of a university faculty or is remote from the university library may not be able to tap whatever library resources exist. The professional's own private library is very likely to contain only a few of the most important resources.

Even when the desired information is found, it is likely that the literature will not provide a complete fit to the problem—close, but not close enough. It will be necessary to take the available information and adjust it or extrapolate it to fit the specifics of the problem. Presumably, the problem parameters are examined and these are compared with the parameters of the material being scanned. If quantitative values are involved, and there are parameter differences between the problem and the available information, perhaps one splits the difference. In any event, logic, intuition, and judgment are involved in adjusting the data.

The amount of effort the user will expend to secure the desired information will vary with the individuals and their need, but is not unlimited. Rouse (1986) and Burns and Vicente (1996) pointed out that there is a cost-effectiveness ratio in information processing. The merely curious users will hardly expend any effort at all, because there is no strong need motivating them. Curiosity is rarely insatiable and the users will exert only that much search effort with which they are comfortable.

More effort will, however, be exerted by someone seeking to solve a problem or to apply the literature to a specific question. However, even this extended amount of effort is not unlimited and the readers/users may at some point, when time is running out and available information sources appear to be unavailing, end the search and "go" with whatever information they have accumulated.

The search may also involve asking one or more colleagues for opinions on the problem. Various methods will be tried, even, for very serious and important problems, the performance of an experiment to secure specifically applicable answers.

The literature may not contain answers to all behavioral problems, but it is possible to increase the probability that the desired information will be found. The strategy requires conceptual researchers to think in advance of the kinds of practical problems the research should help solve. That, however, would require researchers to become more familiar with system development and operation processes. This rarely occurs, although a few system development specialists also perform conceptual research.

The problem-solving strategy is implicit in system-related research (e.g., Which configuration is more efficient?), but not in conceptual research. That is because the experiment (which is preferred in conceptual research) emphasizes the determination of causal factors rather than answers to practical questions.

Many of the actual problems research users are interested in are derived from system development situations in which conceptual researchers are usually uninterested. A possible solution to the problem of applying HF research is to support and publish HF design specialists in research derived from the system development problems they encounter; but this rarely happens. Funding agencies are more interested in conceptual (theoretical) rather than system development questions.

If the concerns of the user are to have some influence on the way research is performed, it becomes necessary to investigate users and the use-factors discussed previously. This requires the development and administration of a questionnaire survey like the one in the following section.

THE USER SURVEY

Methodology

The survey, developed with the aid of T. P. Enderwick, consisted of 29 questions covering the following general themes, the specifics of which will become clearer when detailed findings are presented. These are the same themes discussed previously. The themes include demographic information, frequency and manner of reading the literature, relation of literature use to job-related questions or problems, aspects of the journal papers of greatest interest, usefulness and general user opinion of the literature, the reader's information search process and effort expended in this, and range of reading interests. The survey was mailed to 70 randomly selected members of the test and evaluation (T&E) technical group of the Human Factors and Ergonomics Society (HFES).

Thirty-three responses were received, a return rate of 48%, which contrasts with the 87% return rate for the conceptual structure (CS) survey described in chapter 3. The difference is quite understandable. The chapter 3 survey was mailed to the author's personal acquaintances; the user survey was mailed to a random sample, none of whom was known personally to the author.

Another major difference between the two questionnaires was that in the conceptual structure survey the ideas to be considered by respondents were much more complex than those asked of the users. However, this did not affect the return rate, because a higher return rate would have been expected for surveys with less complex ideas.

Findings

Demographic Information. The sample of professionals in the user survey is very different from that utilized in the CS survey reported in chapter 3. Respondents for the CS survey were much older and more experienced (mean, 29 experience years). The user sample had an experience mean of 16.2 years with a range from 2 to 33.

Of the 33 respondents, 5 designated themselves as T&E specialists, 11 as design specialists, 8 as researchers, 8 as management, and 1 as retired/consultant.

It is often hypothesized that the different functions performed by different types of professionals will produce different responses. A separate analysis was performed of these specialties with a number of indices considered most likely to be sensitive to professional differences. The results of this analysis are discussed later.

Time Spent Reading the Literature. The frequency of reading HF documents is fairly high. When asked how frequently respondents read a HF paper or referred to a handbook, the frequencies were as shown in Table 8.1.

The median percentage time spent reading was 20%, with a range of 0%–4% to 50%. This suggests that at least one half of all professionals spend approximately 20% to 25% of their working time referring to the general HF literature. Of course, it is impossible to know what the interac-

TABLE 8.1
Percentage of Working Time Spent Reading Publications

Percentage of Working Time	*Frequency*
0–4	1
5	2
10	3
15	7
20	5
25	10
30	1
35	0
40	1
45	0
50	3
51–100	0

tion with the literature consisted of, whether it was momentary, longer duration browsing, or a more prolonged information search.

The distribution of frequencies for consulting published government documents was roughly the same, with a median of from 25% to 30%, for those who consulted government publications. Seven, or 21%, did not consult government publications at all. Whether general or government literature, the data suggest the importance of HF publications in the working life of the professional.

Manner in Which the Literature Is Used. One hypothesis entertained at the beginning of the study was that publications were read to answer job-related questions or solve problems. Three, or 9%, *never* read to solve a problem; 7, or 21%, did so *sometimes*; 15, or 45%, did so *usually*; and 8, or 24%, *always* did so. Obviously, the literature is utilized to solve job-related questions or problems, but there is also what can be called a browsing motive that is quite strong.

Twenty-nine, or 90%, reported that they read whatever looked interesting; 26, or 80%, read the literature to keep up with anything novel; 16, or about 50%, read to prepare themselves for research activities. Only 1 respondent indicated reading everything in the field; and only 2, or 6%, read only the HFES publications called *Human Factors* and *Ergonomics in Design*. It is likely that all these reading motivations are interrelated, although all are not operative at the same time. Obviously, respondents have more than one motive for reading HF publications, and the motive operative at any one time depends on external circumstances, such as the need to prepare for a research project.

The usual journal article is subdivided into a number of subsections. It is therefore of some interest to determine how many respondents make use of them: history and background of the research topic: 12, or approximately 40%; study methodology: 16, or 50%; quantitative data: 5, or 15%; results found: 12, or 40%; study conclusions: 19, or approximately 60%; and discussion of results: 6, or 20%. Eight respondents, or 25%, reported finding all sections of the article useful.

Another question directed to the same topic, use of quantitative data in papers, also follows a normal curve, with 63% *sometimes* making use of the data.

The interesting results of this analysis are the relatively low percentage finding the actual data (the "heart" of the study) useful, as well as the relatively few who said the same thing about the discussion section. The data in the study have scientific value to support the conclusions derived from the data, but readers do not make much actual use of those data. The relatively lower percentage of those finding the discussion section useful is also somewhat disheartening.

General Opinion of HF Publications. Respondents were asked about the usefulness of the papers in *Human Factors* and similar journals. These results were: *always*: 1, or 3%; *usually*: 4, or 12%; *sometimes*: 16, or 48%; *rarely*: 10, or 33%; *never*: 2, or 6%. Obviously, journal articles are usually or sometimes useful, but at least one third found them less so.

One finds a similar distribution in responses to a question asking for a general opinion of the HF literature. One found it excellent, 7 found it very good, 18 found it good, 10 found it to be only so-so, and 2 found it poor. Although one third of the respondents had a somewhat negative opinion of the literature, many more had a positive viewpoint. Nevertheless, the editors of any literature supposedly serving professionals, one third of whom have a poor opinion of that literature, should examine their practices.

A common anecdotal criticism of journal papers is that they are remote from real life. Nineteen, or approximately 60%, considered this true; the remainder thought the statement was false. Of course, there is a question of how respondents defined "remote," but however they interpreted the term, it had negative connotations. Although there was a sharp divergence of opinion on this point, the implied criticism should be enough to cause some soul-searching on the part of journal editors.

Behavioral Information Search Process. The search for behavioral data to solve a problem is probably performed in a series of steps. It was therefore of some interest to try to determine the general sequence of the activities involved. Respondents (only 27 answered the question) were asked to list the sequence in which they performed the following. In order of their priority, the activities were:

1. Use of books in an individual's personal library: median rank, either first or second
2. Computer search: median rank, third
3. Reference to a HF handbook (overlaps with 1): median rank, third
4. Use of one's own knowledge: median rank, fourth
5. Discussion of problem with a colleague: median rank, second or third
6. Conduct of an experiment: median rank, sixth

The resulting picture of the search process is somewhat variable, but certain conclusions can be drawn. Conducting an experiment would be the last thing to do, because of the amount of work and expense involved. If an answer cannot be found in the investigators' own knowledge, they may resort to their personal library or a handbook (assuming the latter is not part of that library). If these do not avail, then they may consult a colleague or engage in a computer search.

The one unifying dimension in all this is the amount of effort involved; the professional engages first in the least effortful activity and only progressively engages in more effortful activities, if an answer cannot be found elsewhere. This accords with Zipf's (1965) law of least effort.

One interesting aspect is the relatively high priority given to consultation with colleagues. This accords with Allen's (1977) finding that scientists and engineers often prefer to get their answers from a colleague.

Another item, which dealt specifically with the effort hypothesis, asked whether professionals would be willing to expend a great deal of time and effort to secure an answer to an HF question, if they were certain in advance that an answer existed in the literature. There was a significant split in responses: Thirteen, or 40%, thought the statement was true; the remainder thought that it was false.

Handbooks. A number of questions dealt with handbooks, because these represent an effort to summarize the HF literature and extract the most important information. One question asked if handbooks were useful in the professional's work: 2, or 6%, said *always*; 12, or 36%, said *usually*; 11, or 33%, said *sometimes*; and 5, or 15%, said *rarely*. This suggests that, on the whole, handbooks are viewed favorably by most professionals. Similarly, handbooks are viewed as containing an answer related to a question or problem: *usually*: 13, or 40%; *sometimes*: 15, or 45%; *rarely*: only 4, or 12%.

The ease of using handbooks is a primary factor in their favorable reputation. Twenty-seven, or 82%, felt that this statement was *true*; only 5, or 15%, felt that the statement was *false*. The search of the literature for an appropriate answer to a problem can be a very effortful one; the handbook speeds the process up (at least initially).

Handbooks are easier to use than individual studies, because the handbook creates a context that is more difficult for the individual paper to develop. Because the handbook reviews an entire specialty area (e.g., computer aids), it provides a more comprehensive view than the individual paper. The latter presents only a very brief historical introduction. The handbook refers the reader to the paper, giving that paper some kind of explanatory orientation to assist in the determination of the paper's relevancy.

On the other hand, because handbooks attempt to be encyclopedic, their coverage of individual topics may be too broad brush. There is, in addition, some theory and speculation. The handbook needs to be evaluative as well, pointing out where a line of research has gone wrong or is inadequate. It should point out contradictions, obscurities, and lacunae; but the handbook usually does not do so, perhaps because the scientific code of ethics is too courteous. The individual paper evaluates, because it is required to justify why its research was performed. However, this evaluation deals with only a single line of research.

Does the Literature (Individual Papers and the Research as a Whole) Require Extrapolation or Modification of Its Published Data to Answer Questions? Experience would suggest that this is so. One survey question asked how frequently extrapolation was required: *always*: 4, or 12%; *usually*: 14, or 42%; *sometimes*: 13, or 40%. Only one respondent reported that such adjustments were unnecessary.

Access to the Literature. If individuals do not have access to the literature in the form of libraries or computers, they cannot make effective use of that literature. One question asked whether respondents had *enough* access to journals and handbooks. Twenty-two, or 70%, reported that they had such access; only 10, or 30%, reported in the negative. The operative word in this question was "enough." Everyone has some access, but a significant minority felt that the amount of their access was insufficient.

Concentration on a Specialty. It could be hypothesized that professionals read primarily in the area of their (imagined or real) expertise. One question asked how frequently they read papers *outside* their specialty. Four, or 12%, said *always* or *usually*; 21, or 63%, said this occurred *sometimes*; but 8, or 25%, answered *rarely* or *never*. It seems quite clear that HF professionals are specialists, even in their reading.

Computer Searches. The adequacy of computer searches to provide relevant citations is also important, considering how many professionals make use of this resource to explore the literature. Such searches *usually* (13, or 40%), and *sometimes* (16, or 50%) produce desired results, with only 4, or 12%, reporting that this occurs *rarely*.

Expectations. Does the professional expect to find an answer in the literature to a HF information problem? Eleven, or 33%, felt they *usually* did; 16, or 50%, felt they did *sometimes*; and only 5, or 15%, felt this *rarely*. It appears that HF readers approach their literature searches with some optimism.

Effect of Specialty Function. The most common question asked when there are professional subgroups is whether the various types of HF specialists (i.e., measurement, design, research, management) use the literature differently. A number of survey items were selected as most likely to demonstrate specialty differences. These questions involved the following: reading publications to answer a job-related question or problem; the proportion of time spent reading the general literature on the job; usefulness of journal articles to a professional's work; general opinion about the HF literature; understandability and usefulness of HF papers and handbooks; HF handbooks containing answers; expenditure of time and effort to secure answers to HF questions; need to consult the literature to solve problems; and the realism of journal articles.

In order to provide numerical equivalents for items involving frequency, the frequencies were given the following values: always—5, usually—4, sometimes—3, rarely—2, never—1. True–false statements were given the values: true—2, false—1. The individual frequencies and true–false statements were multiplied by their assigned values, then combined, and divided by the total number of those responding in each subgroup (occasionally some member of the subgroup failed to respond to an item). This gave a fractional value that could be used to compare the subgroups.

Table 8.2 indicates that the four subgroups have small and unequal *N*s. Years of experience were much the same, except for the research subgroup, which had markedly fewer years; whether this is statistically significant cannot be determined because of the small subgroup *N*s. This is also the case for the other items contrasted. The proportions of time spent reading the literature are markedly different for the researcher/design specialist subgroups as contrasted with the T&E and management subgroups.

The other differences among the subgroups seem not to be large enough to represent significant differences among the subgroups. With a larger sample, the differences might be significant, but this is only speculation. A more reasonable explanation might be that even if the subgroups describe themselves in a particular way, each subgroup performs some of the same functions of the other subgroups, so there is sufficient overlap among them to cancel out some performance differences. The same infor-

TABLE 8.2
Response Differences Among Subgroups

Group	*N*	*Experience in Years*	*(1)*	*(4)*	*(7)*	*(8)*	*(14)*	*(23)*	*(26)*	*(27)*	*(28)*
Design Specialist	12	16+	4.0	24%	2.4	2.8	1.7	3.2	1.6	4.0	1.6
T&E	6	15+	3.8	16%	3.5	3.1	2.0	3.3	0.8	3.5	1.8
Management	9	18	3.5	15%	2.5	2.5	1.7	3.5	1.3	3.5	1.3
Research	7	11+	4.3	27%	2.7	2.9	1.7	2.7	1.8	4.1	1.5

Item Descriptions

(1) Use of publications to answer job-related questions.
(4) Proportion of working time spent reading literature.
(7) Usefulness of literature in work.
(8) Opinion of literature.
(14) HF papers/handbooks understandable and useful?
(23) Do handbooks contain answers?
(26) Willingness to expend time/effort to find answers.
(27) Need to consult literature because of questions/problems.
(28) Are HF papers remote from real life?

mation searching and processing operations are probably required, whatever the special functions performed. This is probably true of the CS survey subgroups in chapter 3; professionals who have worked themselves up through the ranks, as it were, have probably performed all the functions descriptive of the various subgroups.

Table 8.2 suggests that there are differences among the professional subgroups, but because of the small *N*s in each subgroup, statistical comparisons are unwarranted. The subgroup designations are generally accepted in the profession, which is why they were used in this study, but they may not be the most significant factors differentiating professionals.

In any event, if the study were repeated with a much larger *N* for each subgroup, some of the differences shown in Table 8.2 might prove significant. But, at the moment, the concept that subgroup interests and backgrounds produce major attitudinal differences must be given a Scots jury verdict of unproven.

CONCLUSIONS

The basic assumption of the user study is that HF research outputs (papers, textbooks, etc.) must be considered to be communication devices. Communication always involves a communicator and a recipient (user), thus efforts to increase the effectiveness of HF research and its communication require that those who communicate (in this case, researchers) must take into account user interests, capabilities, and limitations, provided these are known. Some of those interests are relatively trivial (i.e., curiosity), but others are more significant. One of these more significant interests centers on the applicability of the research outputs to practical work-related problems. Applicability is, after all, one of the goals of HF research.

There is a difference between a *reader* who simply scrutinizes the literature, and a *user*, who uses the literature to solve a work-related problem. There is no question that most HF professionals browse the literature, but a significant proportion also use that literature to solve problems.

In general, professionals are more accepting of the published HF literature than had been supposed. One reason for this may be that professionals are both casual and problem-solvers; in their casual mode (motivated only by curiosity), the literature is accepted in the same way that popular magazines are accepted—that is, for what they are. It is doubtful that reader-professionals question the kind of material they are given, communicated primarily by academics who have the prestige of the university behind them. The basic assumptions behind what is published—the all-consuming importance of the experiment, the necessity of discovering causal factors, the relative insignificance of any other research considerations—are never questioned, so pervasive is the influence of the professional's past training and present experiences. When, however, professionals wish to make use of the literature to answer a question or solve a work-related problem, they may become much more critical of that literature. That may be the reason for the less than unequivocal approval of the literature.

The literature can be criticized on purely conceptual grounds, as in Meister (1999), but the one criticism that cannot be shrugged off and that is purely empirical is the user's opinion of the value of that literature. Despite general approval, an important minority of HF professionals finds the literature mediocre and less than completely useful. Any literature that receives such a report card should cause authors and editors to engage in self-examination.

Whatever the deficiencies of the present study, if professionals accept the notion that research has, in part, a communication function, much more needs to be learned about the recipient of the communication.

So, if anything, what do researchers want to have happen with their work? Certainly they wish their papers to be published and commented on approvingly. This desire may not be overtly verbalized, but it is at the back of the researcher's mind. No one waiting to present a paper at a conference is happy to see only a few blank faces in the presentation room. Much of what the researcher desires—reputation and promotion—depends on the steady flow of well-regarded research. For the researcher, the measurement product, the paper, is all-sufficient and exists primarily for personal gratification.

This is not necessarily a bad thing; many a public good has been achieved through a private motive. One qualification to this must be made. The aforementioned is true primarily for conceptual research published in peer-reviewed journals. System-oriented research has in almost all cases a specific problem to be solved or question to be answered, which serves as its own motive for performing the research. That research will affect the system configuration (if it is under development) or its operation. Most system-related research is, unfortunately, never published in peer-reviewed journals. Enderwick suggested that HF specialists who produce system-related research should try to "sanitize" such research (i.e., eliminate proprietary information, when that is the barrier to publication) and submit the rewritten papers to general journals. Whether peer-reviewed journals would publish such material remains to be seen.

The narcissistic interest of the conceptual researcher means that little or no consideration is given to questions such as: Is the measurement output intelligible to the reader/user? What knowledge or other implications can the reader/user draw from the research/paper? Will the reader/user be able to make any applications of the material presented?

If the researcher performs a conceptual study only to increase the "knowledge store," then the previous questions may be irrelevant and even impertinent. These questions are almost automatic in system-related research, because they are inherent in system-related measurement; the goal of the measurement is problem solution, not adding to the store of knowledge (although the two goals are not incompatible and both may be achieved in the same study).

It could be argued, of course, that narcissistic researchers really do not consider the "store of knowledge" when they attempt to publish a paper. That may be because they assume that, whatever other motivations they have, they automatically add to the store of knowledge whenever a paper is published.

Users have a more than theoretical "knowledge interest" in a piece of research. If individuals are HF design specialists, theoretically they depend on the research to supply design guidelines and specific applications. Nor are only users involved; every casual reader of the literature will eventually, at some time, become a user, because questions/problems inevitably arise that demand answers. Many readers may then be disappointed in the measurement output.

Suggestions for improving this situation have not been unforthcoming. In particular, researchers have been asked to suggest applications of the papers they write. A hypothetical question arises: If researchers can think of *no* practical applications of their work, perhaps that work is irrelevant. "No," says the conceptual researcher, "I contribute to the store of behavioral knowledge, and that is necessary for HF science to progress." Maybe so, but pragmatists may question in the back of their mind whether researchers have perhaps overvalued their contribution.

Art for art's sake is a common concept; the scientific equivalent, science for the sake of science, or the furious pursuit of research for publication purposes, is also a well-established concept, although it is unknown how many conceptual researchers would admit their belief in the concept. Yet it is a reasonable hypothesis that this concept dominates measurement and more particularly the publication process. From the standpoint of a young science like HF, what is the effect of this belief on the discipline? Which, of course, leads inevitably to the question: Where do professionals want HF as a discipline to go and what do they want it to be?

Chapter 9

Final Thoughts

This chapter attempts to pull together some of the main themes from earlier chapters, to expand on a theory of how the human is integrated into system operations, and to make suggestions for solution of some of the problems noted previously.

A LITTLE HISTORY

The concepts discussed here are more understandable in light of a study of the research history of the discipline (Newsletter, 2002a). This study compared papers presented at the HFES annual meetings of 1974 and 1999.

The most obvious conclusion from comparing 1999 with 1974 research is that not much has changed over 25 years. Forty-six percent of the 1974 papers and 55% of the 1999 ones were authored by personnel affiliated with universities. Only a small and decreasing fraction were sponsored by governmental/military agencies (21% in 1974; 9% in 1999) and industry (17% in 1974; 14% in 1999). The experiment remains the most popular measurement methodology, used in 64% of 1974 papers, and 80% of 1999 papers.

The reverse side of the coin is that use of nonexperimental methods (e.g., questionnaires, surveys) decreased from 34% in 1974 to 10% in 1999. At the same time, research directed at system design/development was only an insignificant fraction of all *Proceedings* research (10% in 1974, 6% in 1999), although obviously industry continues to perform substantial HF research (although unpublished) in support of its activities. If the search for causal factors of phenomena (significance of differences between treatment conditions) is a characteristic of the university-centered experimental approach, it would be expected that much HF research would reflect the same interests; and so it does (between 30% and 40% of papers in the two time periods).

Other continuing trends were manifested. Subjective data were gathered from questionnaires and debriefings, for example, in both time periods (45% in 1974, 73% in 1999), but those merely supported objective data. Subjective data did not provide any results that controverted or illuminated the objective data. The development of theories and models (supposedly one of the reasons for research) was not common (19% in 1974, 24% in 1999). The great majority of researchers suggested few applications of their study results (23% in 1974, 34% in 1999), and, even when they did, most recommendations were simply airy abstractions.

Real life often presents personnel with indeterminate (i.e., uncertain) situations involving the selection of alternative strategies. In contrast, in both time periods, most studies involved highly determinate test situations, meaning that stimuli were unambiguous and clear-cut (68% in 1974, 58% in 1999). It is possible that the need for relatively simple, concrete measurement situations to permit use of controlled conditions encouraged this tendency.

There were, however, some marked contrasts. In 1974, the ratio of empirical to nonempirical papers (tutorial/case history/analysis) was only 30% to 70%. In 1999, this ratio was completely reversed. The ratio of system-oriented papers (those involving more molar system elements) decreased from 1974 (38%) to 1999 (24%). This suggests that HF research has tended to concentrate on the more molecular system elements (workstations, interfaces). This may be because more molecular equipment elements are easier to utilize in a highly controlled experimental measurement framework. There was much less emphasis in 1999 on "nuts and bolts" research (e.g., console dimensions, anthropometry, the effects of environment on personnel performance) than in 1974.

The picture that emerges from this analysis is of a very academically oriented research effort based on a few implicit (and hence never discussed) assumptions: The experiment is the preeminent measurement method; research should be directed toward discovering causal factors; and everything else (e.g., the development of a quantitative database and of a quantitative predictive technology; the human–machine system as a source of fundamental HF problems) is unimportant and can be ignored.

The research based on these assumptions emphasizes the human (rather than the system) response to technological stimuli, which means that it is very similar to psychological research. Most conceptual, basic, nonapplied research (choose the term the reader prefers) is performed in highly controlled, laboratory-type test situations, with little concern for human performance in the operational environment (OE). Researchers appear to be largely unconcerned with the application of their work to system development; in consequence, their results have only a limited relevance to actual use situations. The question to be answered is: What factors are responsible for this type of research?

To discover the most common research interests of HF professionals, an analysis was made (Meister, 2003) of the themes of papers presented at annual meetings of the Human Factors Society during four periods (1959–1969, 1970–1975, 1977–1980, and 1981–1985). The research themes exhibited both continuity and discontinuity, the latter influenced by governmental funding of research on systems like automobiles, space flight, and nuclear power. The frequency of system-related HF research waxed and waned with contracts. Although there was a wide range of themes (170 in 1985), most excited only one or two studies. Readers may ask how much can be learned from such widely scattered efforts. Most frequently studied themes were of system-related behaviors like driving in automobiles instead of fundamental functions like perception. This suggests a different research orientation from that of psychology. Perhaps the most interesting finding was that there were almost no studies involving fundamental HF assumptions and what these meant for HF research. It was as if the discipline had been born like Athena, full grown from the head of Zeus. This has had unfortunate implications for HF research.

HUMAN FACTORS TENSIONS

Underlying HF in general and its measurement in particular, are a set of tensions between apparent incompatibilities. These tensions help to create the dynamic for HF and its measurement. This dynamic, along with HF goals, is what makes a system (in this case the HF discipline) move in one direction or another.

Such tensions exist in every discipline, but the nature of the tension differs in accordance with the nature of the discipline. In HF, the tensions are particularly severe, because the discipline contains dichotomies (e.g., between the behavioral and physical domains, between the individual as an individual and the individual as part of a system, between research and application) that lead to these tensions. These tensions may appear to be only the author's intellectual conceit, but they have a significant effect on how the discipline performs. The tensions exist below the surface of HF activity (professionals generally do not think of them), because customary measurement practices overlay and conceal them. These tensions present tough problems for the HF community, which is why they have generally been ignored.

The most fundamental tension is that between the two major HF purposes. On the one hand, there is the felt need to explain the occurrence of HF phenomena (as a result of which there is conceptual research and the derivation of general explanatory principles). On the other hand, there is the need to transform and to incorporate those principles into human–machine system development. There is no necessary incompatibility between these two purposes, but the apparent dichotomy leads to an undue overemphasis on one or the other, particularly on so-called basic research. The overvaluation of "knowledge" as a scientific phenomenon, as opposed to the application of that knowledge to actual phenomena, has produced a tendency to ignore system problems.

Related to the preceding is the dichotomy between the *physical* and *behavioral* domains, which are intermingled in HF. HF, as a behavioral science, seeks to apply behavioral data and solutions to a physical domain, that of technology in general and the human–machine system in particular. The struggle to translate one into the other underlies the entire HF enterprise, as may be seen in the effort to transform behavioral research conclusions into design guidelines for physical systems.

The tension between the physical and behavioral domains is reflected primarily in the difficulty of applying behavioral principles and data to system design problems. Design guideline principles developed from research studies provide general guidance, but many of these are only qualitative. This may result from the difficulty of applying the general to the particular, but it goes beyond that. How does the researcher apply behavioral data? The data may be applied to the human operator, which means predicting quantitatively what that human will do in operating the machine (e.g., in the form of errors). The HF community has been disinclined to try to predict human performance quantitatively (for more detail on this, see Meister, 1999).

This tension creates problems in simulation, but more fundamentally it reflects an uncertainty as to what constitutes "reality" for HF. It is assumed that HF research results are relevant to and generalize to that reality, but there is no way of actually determining this; consequently, professionals merely assume generalization of the research and let it go at that. They also lack an operational definition of what research generalization actually consists.

Simulation involves reproducing the characteristics of real-world operations, on the assumption that this simulation will produce real-world operator behaviors. Simulation may be either physical (in the form of simulators) or behavioral (human performance models). Physical simulation is largely limited by cost; behavioral simulation is much less expensive, but much more difficult to achieve. The major problem of implementing models is that one needs real-world data inputs, and there is a general reluctance to gather data for the sake of making such inputs. (Data to prove/disprove hypotheses is a different matter.)

Another tension is between what is considered *basic* research, with its "higher" intellectual status but its correlated problem of applicability and utilization, and what is considered *applied* research, which may have somewhat "lower" intellectual status but more immediate usefulness. The dichotomy between basic and applied research is accompanied by snobbism: Basic research is supposedly more academic, more sophisticated, more intellectual, more fundamental than applied (defining applied as whatever is not basic); this provides a convenient excuse for the researcher to ignore a variety of real-world problems that are difficult to attack. The author's definition of what is basic (e.g., the transformation of behavioral phenomena into physical equivalents, which is the essence of making behavioral inputs to system development) would include many things more conventional professionals would reject as applied. Again, this dredges up the same difficulty discovered in chapter 3: the inability or reluctance of the HF community to define fundamental constructs. Basic research in most behavioral journals is primarily operator centered. A rule of thumb in the academic community is that any research that involves equipment as a major element in the research is applied. Needless to say, this discussion does not accept this proposition.

There is the more mundane tension between the presumed neutrality of the conceptual research effort, dedicated to the production of knowledge, and the need to secure *funding*. Funding agencies have their own agenda, which often conflicts with priorities of scientific research. Funding is always limited, so there must be a choice among the research projects that will be funded. That selection is not unaffected by the biases of the funding representatives, which reflect the general attitudes of the academic community. It is striking, however, that the assumed objectivity of the search for scientific knowledge is mediated by the extreme subjectivity with which funding decisions are made.

A major tension exists between the concept of humans as individuals and their role as an element of the human–machine system. Part of the problem is that the individual plays both roles at the same time and the roles are somewhat incongruous. In the former, humans are considered to function independent of the system; in the latter, they are forced by the system to perform to system requirements, which may conflict with personal tendencies. In their independent role, humans are best studied using psychological principles (i.e., as simply a respondent to stimuli) following the S–O–R paradigm. In the system role, humans must be studied in accordance with

HF requirements, which view humans as a proactive/interactive element of the system. As a consequence, sometimes what is intended to be a HF study turns out to be a psychological study, and thus lacks relevance for HF.

The difficulty of distinguishing between humans as individuals (within their skin, as it were), and humans as elements of the system means that research on the latter is often ignored. There is also the problem of including the system in test situations. The influence of the S–O–R paradigm in HF also tends to emphasize the human as the center of the measurement, whereas system-related research downgrades the human from that center and makes the system the center. (The astronomical parallel—Ptolemaic vs. Copernican—comes to mind.) Research on the human as the center of the study is logically the province of psychology, to which the system is foreign. The system in a study is not merely a matter of presenting subjects with an equipment or equipment components. The system in a study must have a goal, a mission, and the interaction of subsystems. Without these, a study cannot be considered system related.

As a result, a tension exists between the influence exerted by the predecessor discipline, psychology, and the questions, problems, and methods inherent in the successor discipline, HF. The continued use of the S–O–R paradigm in HF often creates problems for the HF professional of determining *what* should be studied and *how*. As a result, much HF research is focused on the human, as if the system did not exist; such research is psychological, rather than HF, in nature.

Many HF professionals are not aware of the impact of the predecessor discipline on their thinking and research methodology, an impact felt primarily in the selection of research topics. Also involved in the psychological tradition is the overemphasis on the experiment as the preferred research methodology and the reluctance of many professionals to include the system in their studies. Many HF professionals think of themselves as psychologists (Hendrick, 1996), which narrows their view of the purposes of HF and potential alternatives in the discipline.

Another form of tension is between the quantitative fact or *datum* and the *fact* as interpreted in the form of *conclusion*, speculation, or theory. Data can be interpreted only in the context of the measurement circumstances from which they were derived. This means that a particular human performance is laboratory-type performance, not performance as it would be in the real-world, operational context. However, the thrust of all measurement is to generalize from the specific test situation to a wider use context. Data are theoretically not influenced by human biases, but without interpretation by humans, as they are seen in conclusions, speculations, and theory, the data lack meaning. To apply meaning to data, however, inevitably involves human biases.

The transformation of data into a conclusion is also the transformation of the *specific* (data) into the *general* (principles). When the study is an experiment, the nature of the transformation is determined by the hypotheses being tested by the experiment. For example, a specific type of human performance is determined by Factor X, shown to be statistically significant. Unfortunately, the importance of Factor X relative to any other factors influencing that performance is not indicated by the statistical significance level; the latter merely demonstrates that, under the test conditions producing the data, the results related to Factor X were not produced by chance. The researcher's judgment of the importance of the Factor X conclusion is somewhat biased by a predisposition to magnify that importance. This magnification is created by the words used by the study author, particularly their connotations.

It is often difficult to distinguish between the fact as it emerges from test records and the fact commingled with speculation and theory. The unfortunate result is that professionals often do not know when they are dealing with "hard" fact or "soft" speculation or indeed what hard facts exist.

Beyond these more specific tensions there is a general tension between the *effort* needed to solve the problems described in this text, and the lack of *will* to mount that effort. Some professionals are acutely aware of these problems, but are unable as individuals to sustain the effort required to attack them.

What do these tensions mean to the discipline? In HF, tensions result at least in part from an overemphasis on one or the other element of a dichotomy. This emphasis has been tilted to the behavioral over the physical and to conceptual research over system-oriented needs.

Tensions can be positive as well as negative. To have any effect on a discipline, they must first be recognized and applied to the problems to be solved. If professionals deliberately try to utilize these tensions as a motivating and directing drive for their research, the results may be very worthwhile. The purpose of that research would then include the following motivations:

1. To explore the two domains of the physical and behavioral as they relate to human performance, and to overcome the barriers between them.
2. To make conceptual research accommodate (apply) itself to system development, which HF does only incompletely.
3. To accommodate the various purposes of HF, which require it to aid system development (i.e., to be proactive) as well as to explain human performance phenomena (a post facto attitude). This produces divergences in HF research.

Although the tensions can be used as a driving need for change, it is not clear that HF researchers are aware of this possibility. If these tensions rarely influence the average HF professional, it is because the latter are responsive mostly to the minutiae of daily work, the procedures implicit in customary measurement practices, which tend to "narcotize" the tensions.

Another reason why these tensions are submerged below the level of the professional's consciousness is, in part, because they are rarely discussed—not in symposia held by HF societies, because the societies, like the professionals whose behaviors they reflect, are concerned mostly with empirical testing details and lack patience with more molar analysis. More important, however, is the lack of discussion of these topics by university graduate departments, which are absorbed with experimental designs and statistics, and care little for what these designs and statistics *mean*. Student professionals can hardly be expected to develop these ideas on their own and, when they enter the real world, they are constrained by work limitations.

A THEORY OF HUMAN–SYSTEM FUNCTIONING

Much was made in previous chapters of the system and the integration of the human into the system. There has, however, been no formal theory of how this integration occurs. The following is a very preliminary attempt at such a theory.

A theory should attempt to answer the following questions:

1. How is it possible to differentiate between human and machine contributions to system effectiveness?
2. What discernible differences in *types* of systems produce significant differences in personnel performance?
3. What is the effect of human performance at subordinate system levels on terminal system success/failure?
4. How is the effect of human performance (e.g., error) at these subordinate system levels (e.g., the workstation) manifested or produced at higher system (e.g., command) levels?

Any theory attempting to answer these questions must make use of data previously gathered on real-world systems, or at least highly realistic simulations. The system theory discussed later does not answer these questions, because the empirical data infrastructure needed for the theory does not yet exist. The theory is at best an *outline* of what a truly comprehensive human–system theory would be.

Present HF theories (e.g., the multiple resources theory, Wickens, 1991) are actually only psychological theories in HF disguise, because they focus on the human as essentially independent of the systems in which humans function. For example, theories of workload and stress conceptualize the human as merely responding to sources of stimuli; no system context is included in these theories.

The human functions *concurrently* in two ways: as an individual responding to stimuli of various types and as an individual responding to requirements imposed by human–machine systems of a social, economic, or cultural nature (e.g., school, work, military, entertainment, communication, or transportation).

All these systems utilize technology to one degree or another, which is why even religious systems, such as a church, can be called human–machine systems, while at the same time being religious systems. For example, the church uses telephones, computer systems, automobiles, and so forth.

It cannot be emphasized sufficiently that the study of the human as a system element is the province of HF. It is assumed that the principles that animate the functioning of the human as a system element are different from (although related to) those that apply to humans in their individual personas.

The difference between these two roles may not be clear to some. Because the individual functions in both, it may be somewhat difficult to differentiate HF from psychological concerns. In both roles, the human senses stimuli and responds to these, so the S–O–R paradigm derived from psychology applies in both situations, although only partially in the HF situation.

The imposition of an external task and goal changes the human's performance as an individual. Soldiers in a trench who hear the command to move up and attack would not do so if *only* their internal stimuli and responses were functioning. The nature of the task and goals thus becomes a matter of HF study.

Because of its system context, professionals cannot study human performance as they would if the system did not exist. The researcher must take system/subsystem interactions, its goals, and its tasks, into account in explaining the human's performance. That is why conceptual research, which ordinarily does not include a system in its measurement procedures, presents a problem. The fact that the human performance it records is isolated from any system context makes application of its results difficult.

Principles of Human–System Interaction

The human, as a system element, is controlled by the rules established by that system. It does not matter that the system was originally designed and constructed by other humans who specified the organizational rules by which all those who become part of the system must now function. If those who developed the system are presently those who control the system, then they are also part of and controlled by these rules. For example, CEOs of industrial companies can be dismissed if the system decides to dispense with their services. It is important to remember that because humans create systems, a necessary HF goal is to study how they design systems and to aid in their construction by providing HF inputs.

There are many systems. An individual need not be part of every system, although all humans are part of some systems. For example, individuals need not be members of the military, but (except in rare instances) all children must enroll in school. Once part of a system, voluntarily or not, humans are controlled by that system. The system tells them what must be done, presents stimuli, evaluates responses, provides rewards (e.g., salaries), and so on. Individuals may dislike the notion of systems that control humans, but that is a fact of life.

However, the human is not completely controlled by the system. Where humans are volunteers, they can usually opt out of the system: run away from school, quit a job, decide not to view a particular motion picture, and so on. However, in most cases, the human, once entered in the system, remains part of the system. One of the assumptions underlying the human–system theory is that how well that human performs as a system element determines in some measure how well the system as a whole performs. If that relation did not exist, then there would be no need for a human–system theory, or even for a HF discipline. The mutual relation between the system and the human is known as *symbiosis*, a phenomenon common in nature. The human creates the system; the system nurtures and controls the human. As an academic discipline (because it is that too) HF is responsible for studying this symbiotic relation.

It follows, then, that the criterion of human performance effectiveness as a system element is how well the parent system functions, and how much personnel contribute to system success. Individual performance (independent of the system) can be described by error/time scores, subjective measures, and by wages; performance of system personnel (considered as a whole) can be described only by how well the system functions, by whatever performance criteria the system has established for those personnel.

The system, like any other source of stimuli, makes certain performance demands on individuals, and humans must draw on their resources (cognitive, physical, etc.) to satisfy those demands. This is the paradigm of workload that plays such a large part in HF research. Workload may seem to be a peculiarly individual phenomenon, but it must also be conceptualized as something that happens to the workforce as a whole. It is that latter effect in terms of system personnel as a whole, which is as important as the workload felt by the individual—at least from the standpoint of HF.

System stimuli are both molecular (e.g., a scale reading on a meter) and molar (requirements for the performance of system functions). Molecular stimuli lack meaning unless they can be related to the larger system context. Humans respond first to the molecular stimuli (e.g., the scale reading), but in order to make sense of these and to implement the system function, they must relate the stimuli to the functions, tasks, and goal of the entire system or subsystem. When personnel are well trained, the relation may be almost unconscious.

The operator transforms these molecular stimuli into more meaningful system stimuli by applying learned *information* to the former. An essential aspect of that information is the relation of the molecular stimuli to system functions. A meter reading describing boiler pressure means nothing in itself; the operator's knowledge (that boiler pressure relates to water flow between two nuclear reactor subsystems, and that water flow has some relation to core temperature) enables the operator to perform the system function of adjusting water pressure or shutting the boiler down.

The system is represented not only by molecular stimuli like meter readings, but also by molar characteristics, which can be called *attributes*, like complexity, transparency, and flexibility. Attributes incorporate many molecular stimuli, but the former refer to and characterize the system as a whole. These attributes provide the *context* within which personnel respond to the molecular stimuli the attributes include.

Most important is the system goal and its accomplishment, which transcends and overlays any individual goals. This means that (when the system begins to function) the individual's goal is absorbed by the system goal so that operators lose track of their personal goal. In the process, the system goal subsumes the individual's goal and changes it; the individual's goal is now to implement the system goal. If the system goal is not achieved, neither is the individual goal; this explains why the human feels downcast when, for example, a sports team, to which the human is passionately attached, loses. The individual's goal is to perform effectively, but effectiveness is defined by system success.

The operator in the S–O–R paradigm (the *O* in S–O–R) is in HF terms not the individual operator, but all the system personnel who contribute to system functioning. What individuals do is important to them, but is relatively unimportant unless it implements a system function and affects system success or failure. The *R* in the system S–O–R is represented by the accomplishment of the system goal.

Of the system attributes mentioned previously, complexity is the most important, because it is the primary source of the demand imposed by the system on personnel. It, therefore, has a negative effect on human performance. Transparency (the display of internal functions to the operator) and flexibility (the operator's ability to switch from one response mode to another) are assumed to counteract, to some extent, the effects of complexity. There may well be other system attributes, like type of organization, which are also important.

Complexity can be defined in several ways. In equipment terms it is the number of components, number of equipment interactions, and so on, but these are of relatively little interest to the HF professional unless their effects are displayed on a human–machine interface that must be operated. In human terms, complexity is represented by, for example, the number of interface tasks to be performed and the number of critical events to be monitored or decided on. This is human complexity viewed objectively. That complexity is encapsulated in the interface, because (except for maintenance tasks) this is where operators direct most of their attention. Human complexity can also be represented in subjective, cognitive terms (e.g., the operator's ability to understand system interrelations or symptomatology, or by the operator's expressed difficulty in performing a task).

The physical mechanisms needed to implement system functions may require a certain degree of complexity, but designers are able, within limits, to select and arrange those mechanisms to reduce unnecessary complexity. The HF design specialist has the responsibility to assist the engineer in this respect by providing information about how humans respond to complexity elements.

The reader can ask what the impact of such a theory might be on HF research, if the theory organized that research. What would change—if the theory were implemented in research—would be what professionals would study, and how research questions would be studied. The following study topics are only illustrative:

1. The effect of system characteristics like complexity, organization, and so forth, on personnel performance; this would first require the development of methods of measuring these characteristics.
2. The mechanisms by which personnel contribute to system goal accomplishment (e.g., how individuals and personnel as a whole implement system functions); how personnel performance at various subsystem levels interacts; how to quantify the human contribution to system performance.
3. How engineers design systems and how HF data inputs can assist in this process.
4. Very generally, how HF can translate behavioral principles and data into physical mechanisms.

Topic 4, which ought to be the foundation of HF research, is barely considered in present HF studies, because most researchers do not think in transformational terms.

Methods of answering system questions will deviate somewhat from those presently employed in most HF research. If professionals include the system in their research, they may have to deal with actual systems in the operational environment or create realistic representations (simulations) of such systems. The simulations may be physical (e.g., an aircraft simulator) or symbolic (computerized human performance models).

RECOMMENDATIONS

The picture of HF measurement painted in this book suggests certain deficiencies: overemphasis on humans as opposed to the system context in which they function; the lack of a formal, systematic, quantitative database (despite many HF texts and handbook compilations); the inability to predict human performance, in part because that quantitative database is lacking; the overemphasis on the experiment and the conceptual study, so that other measurement methods are neglected; the overemphasis on the specialty area and the funding agency as the factors that determine the themes of HF research.

The deficiencies already cited are not absolute errors or failures, but rather the result of an excess of attention paid to one aspect of measurement rather than another. Because these are preoccupations, it is possible to adjust them. Although no one would, for example, recommend abandonment of the experiment and the conceptual study, these can be balanced by other measurement approaches.

The following contains a number of recommendations. These recommendations pertain as much to the discipline as a whole as to measurement in particular, because the two are inextricably bound together.

The Need for Self-Examination

HF professionals need to examine the state of their discipline on a discipline-wide basis. This means they should make a deliberate effort to answer the following questions:

1. What do they actually *know* (as opposed to what they think they know, or speculate, or theorize) about HF variables and functions?
2. What data does HF need most to have?
3. What research needs to be done to secure those data?

Most HF professionals would argue that this self-examination does occur on an individual paper basis, in the form of historical reviews of past research; but all such reviews are based on specific research problems (e.g., what needs to be known about icons), and do not consider the needs of the overall discipline.

The essential word is *need*: What do the discipline and the specialty areas (e.g., aerospace) need to know and how can these needs be combined? Answers to these questions must be based on the aims of the discipline and the specialty area: to assist in system development, to predict future human performance, and to explain how people perform. It is necessary to begin with the goal, not with whatever research has been performed up until now. It is neither feasible nor desirable to discard previous research, but research can be examined in terms of whatever goals are accepted. Researchers have to know where they are going before initiating a process, or setting out on a voyage.

It is hardly to be expected that any single professional will establish the goals of the discipline. Moreover, the goals of the discipline are made more concrete when they are referred to the content of the specialty area. It is, however, feasible for the representatives of the discipline, the HF societies, for example, to give themselves the task, by establishing commissions (that much overused mechanism, but nevertheless useful) of experts in each major specialty area to consider the questions raised in this book, and to develop means of answering these questions. The various HF societies are important, not for what they presently do (which presently is not much), but for what they could do, if they took the lead in the self-examination process.

The starting point of any such self-examination is the development of a consensus on the goals of the discipline. It will require vigorous debate to arrive at such a common understanding. The output of such self-examination could be position papers, proceedings of symposia, or even a text. The voice of average HF professionals ought to be heard in this debate, because, as a whole, they are—whether they like it or not—the discipline. Self-examination is not a one-shot affair. The debate must be a continuing one, because over time new questions will arise and must be examined.

Another locus for the self-examination process is the university, which trains novice professionals and inculcates many of the tendencies reviewed in this book. A review of contemporary textbooks used in university HF courses (e.g., R. W. Proctor & Van Zandt, 1994; Wickens, 1992) is a vain search for any consideration of the questions raised in this book. There is a brief and usually misleading discussion of validity, a note on generalization of results, and then the reader passes almost immediately to a short discussion of measurement and a much more detailed dissertation of experimental design and statistics. The concerns raised in this book are almost never noted, although they may appear fleetingly in graduate seminars.

However, the academic community feels about the validity and importance of the questions raised in this book, there should at least be an interest in refuting the points of view (admittedly nontraditional) presented by the author. If academics have a conceptual structure different from the author's, then why not express it to their students? (Few HF texts begin, as this one does, with a listing of the concepts that underlie the principles discussed.) The classroom is an ideal forum to raise these notions and it is possible that some of the students may develop a somewhat fresher point of view about HF than they have presently.

Measurement Strategies

One of the questions that might be discussed in any debate on HF measurement is an overall strategy for that measurement. Every professional knows that the present research strategy is to search for causal factors, to hypothesize variables, and to perform a conceptual study requiring an experiment.

HF research can be performed by asking questions other than those dealing with causal factors. One alternative might be to assume that the major function of HF research is to assist in solving system development and operational problems. Researchers might begin by uncovering behavioral problems that arise in system development and perform research to discover the factors causing these problems. System-related research is very close to this orientation, although the focus of most system-related research is fairly narrow, restricted to the individual problem to be solved. However, human–machine systems, being molar entities, do lend themselves to the study of general principles. Experimentation might or might not be the preferred method of securing answers to the problems they present.

Another way of performing research is also problem oriented, but begins with the question asked earlier: What questions do researchers need to address for which they do not have answers? The sources of these needs to be satisfied by research could be the same elements presently utilized: logic, experience resulting from observation, available theory, and previous research. The paradigm is: Need leads to research.

In fact, all the preceding alternatives are in use to some extent, although only the causal factors approach is accepted as correct by most researchers. Those who adopt the causal factors approach assume that knowledge of the effects of variables will automatically lead to the data needed by the discipline. The correctness of this assumption has never been tested, but if it were, it might be found that experimentation does not necessarily produce all or even most of the data needed by the discipline.

What distinguishes these alternatives is the source of the research motivation. In every case, a problem, conceptual or empirical, initiates the research; the only difference is how the problem arises. It may derive from any or all of the sources previously noted. Most conceptual research derives not from an observed problem, but from logic, available theory, and previous research results. System-related research stems from actually experienced problems. The approach involving the need to fill gaps in knowledge emphasizes all the sources of the need. The difference among the approaches is a matter of degree. The conventional approach to conceptual research ignores to a certain extent developmental and operational problems (as opposed to previous research). The problem-oriented approach is likely to concentrate on empirical experience. The "needs" approach is the one that appears to be most inclusive of the sources motivating research.

HF researchers do consider what is needed and the gaps in knowledge that must be filled. These gaps, however, involve only the individual specialties (e.g., aerospace) and specific problems within each area. The needs of the discipline as a whole are largely ignored.

Of course, this consideration must begin with specialty areas, because they exist, but many of the needs in each specialty are the same. For example, when each technical interest/specialty area in the Human Factors and Ergonomics Society was asked, in 1995, to list their major research needs, the similarity among the lists supplied was striking (Newsletter, 2002b).

CONCLUSIONS

Much of what has been presented in this book is controversial, which is why the previous recommendations require, as a preliminary, discussion of the questions raised and the solutions suggested. Such a discussion, if it occurs, would be a significant step toward the implementation of the recommendations.

Before any of this can occur, however, an effort of *will* (discussed earlier) is necessary. As everyone knows, it is easier to accept the status quo as a given; change involves pain. Consequently, there is no great expectation that the discussion called for will in fact occur. At most, a few individuals may be convinced, not of the essential rightness of the ideas expressed here, but of the need to think about and discuss those ideas.

References

Advisory Group for Aerospace Research and Development (AGARD). (1989). Human performance assessment methods (Rep. No. AGARD-AG-308). Neuilly-sur-Seine, France: Author.

Allen, J. J. (1977). *Managing the flow of technology.* Cambridge, MA: MIT Press.

Babbitt, B. A., & Nystrom, C. O. (1989a). *Questionnaire construction manual.* Alexandria, VA: Army Research Institute for the Behavioral and Social Sciences.

Babbitt, B. A., & Nystrom, C. O. (1989b). *Questionnaire construction manual annex: Questionnaires: literature survey and bibliography.* Alexandria, VA: Army Research Institute for the Behavioral and Social Sciences.

Bedny, G. Z., & Meister, D. (1997). *The Russian theory of activity: Current applications to design and learning.* Mahwah, NJ: Lawrence Erlbaum Associates.

Berliner, D. C., Angell, D., & Shearer, J. W. (1964). Behaviors, measures, and instruments for performing evaluation in simulated environments. In *Proceedings, Symposium on Quantification of Human Performance.* Albuquerque, NM: University of New Mexico.

Boff, K. R., & Lincoln, J. E. (1988). *Engineering data compendium: Human perception and performance.* Dayton, OH: Wright-Patterson AFB.

Boldovici, J. A., Bessemer, D. W., & Bolton, A. E. (2002). *The elements of training evaluation.* Alexandria, VA: U.S. Army Research Institute.

Bortulussi, M. R., & Vidulich, M. A. (1991). An evaluation of strategic behaviors in a high fidelity simulated flight task. *Proceedings of the 6th International Symposium on Aviation Psychology, 2,* 1101–1106.

Box, G. E. P., & Draper, N. R. (1987). *Empirical model building and response surfaces.* New York: Wiley.

Box, G. E. P., Hunter, W. G., & Hunter, J. S. (1978). *Statistics for experimenters: An introduction to design, data analysis, and model building.* New York: Wiley.

Boyett, J. H., & Conn, H. P. (1988). Developing white-collar performance measures. *National Productivity Review,* Summer, 209–218.

Burns, C. M., & Vicente, K. J. (1996). Judgments about the value and cost of human factors information in design. *Information Processing & Management, 32,* 259–271.

Burns, C. M., Vicente, K. J., Christoffersen, K., & Pawlak, W. S. (1997). Towards viable, useful and usable human factors design guidance. *Applied Ergonomics, 28,* 311–322.

Chapanis, A., Garner, W. R., & Morgan, C. T. (1949). *Applied experimental psychology: Human factors in engineering design.* New York: Wiley.

Charlton, S. G. (1992). Establishing human factors criteria for space control systems. *Human Factors, 34,* 485–501.

Chiapetti, C. F. (1994). *Evaluation of the Haworth–Newman avionics display readability scale.* Unpublished thesis, Naval Postgraduate School, Monterey, CA.

Cooper, G. E., & Harper, R. P., Jr. (1969). *The use of pilot rating in the evaluation of aircraft handling qualities* (Rep. No. NASA TN-D-5153). Moffett Field, CA: NASA-Ames Research Center.

Czaja, S. J. (1996). Aging and the acquisition of computer skills. In W. A. Rogers, A. D. Fisk, & N. Walker (Eds.), *Aging and skilled performance; Advances in theory and application* (pp. 201–220). San Diego, CA: Academic Press.

Czaja, S. J. (1997). Using technologies to aid the performance of home tasks. In A. D. Fisk & W. A. Rogers (Eds.), *Handbook of human factors and the older adult* (pp. 311–334). San Diego, CA: Academic Press.

Davies, J. M. (1998). *A designer's guide to human performance models* (AGARD Advisory Report 356). NATO Research Group paper, Neuilly-sur-Seine, Cedex, France.

Eberts, R. (1987). Internal models, tracking strategies and dual task performance. *Human Factors, 29,* 407–420.

Endsley, M. (1995). Towards a theory of situation awareness. *Human Factors, 37,* 32–64.

Flanagan, J. C. (1954). The critical incident technique. *Psychological Bulletin, 51,* 327–358.

Fleishman, E. A., & Quaintance, M. K. (1984). *Taxonomies of human performance: The description of human tasks.* Orlando, FL: Academic Press.

Gawron, V. J. (2000). *Human performance measures handbook.* Mahwah, NJ: Lawrence Erlbaum Associates.

Gawron, V. J., Schifflett, S. G., et al. (1989). Measures of in-flight workload. In R. S. Jensen (Ed.), *Aviation psychology.* London: Gower.

Giffen, W. C., & Rockwell, T. H. (1984). Computer-aided testing of pilot response to critical in-flight events. *Human Factors, 26,* 573–581.

Green, D. M., & Swets, J. A. (1966). *Signal detection theory and psychophysics.* New York: Wiley.

Hancock, P. A., & Caird, J. K. (1993). Experimental evaluation of a model of mental workload. *Human Factors, 35,* 313–319.

Hart, S. G., & Staveland, L. E. (1988). Development of NASA-TLX (Task Load Index): Results of empirical and theoretical research. In P. A. Hancock & N. Meshkati (Eds.), *Human mental workload* (pp. 139–183). Amsterdam: North Holland.

Hendrick, H. W. (1996). All member survey: Preliminary results. *HFES Bulletin, 36*(6), 1, 4–6.

Hoffman, R. R., & Woods, D. D. (2000). Studying cognitive systems in context: Preface to the special section. *Human Factors, 42,* 1–7.

Kennedy, R. S. (1985). *A portable battery for objective, nonobtrusive measures of human performance* (Rep. No. NASA-CR-17186). Pasadena, CA. Jet Propulsion Laboratory.

Kirwan, B. (1994). *A guide to practical human reliability assessment.* London: Taylor & Francis.

Lanzetta, T. M., Dember, W. N., Warm, J. S., & Berch, D. B. (1987). Effects of task type and stimulus heterogeneity on the event rate function in sustained attention. *Human Factors, 29,* 625–633.

Laughery, K. R., Jr., & Corker, K. (1997). Computer modeling and simulation. In G. Salvendy (Ed.), *Handbook of human factors and ergonomics* (pp. 1378–1408). New York: Wiley.

Laughery, K. R., Corker, K., Campbell, G., & Dahn, D. (2003). *Modeling and simulation of human performance in system design.* Unpublished manuscript.

Maltz, M., & Shinar, D. (1999). Eye movements of younger and older drivers. *Human Factors, 41,* 15–25.

McDougall, S. J. P., Curry, M. B., & de Bruijn, O. (1998). Understanding what makes icons effective: How subjective ratings can inform design. In M. A. Hanson, *Contemporary ergonomics, 1998* (pp. 285–289). London: Taylor & Francis.

Meister, D. (1971). *Human factors: Theory and practice.* New York: Wiley.

Meister, D. (1985). *Behavioral analysis and measurement methods.* New York: Wiley.

Meister, D. (1998). Relevance and representativeness in HFE research. *Proceedings of the Human Factors and Ergonomics Society Annual Meeting,* 675–678.

Meister, D. (1999). *The history of human factors and ergonomics.* Mahwah, NJ: Lawrence Erlbaum Associates.

Meister, D. (2003). Empirical measurement of HF research themes. *Proceedings of annual meeting of the Human Factors and Ergonomics Society.*

Meister, D., & Enderwick, T. P. (2001). *Human factors in design, development, and testing.* Mahwah, NJ: Lawrence Erlbaum Associates.

Mertens, H. W., & Collins, W. E. (1986). The effects of age, sleep deprivation, and altitude on complex performance. *Human Factors, 28,* 541–551.

Meyers, R. H. (1990). *Classical and modern regression with applications* (2nd ed.). Boston: PWS-Kent.

MicroAnalysis and Design. (1993). *MicroSaint for Windows user's guide.* Boulder, CO: Author.

Montgomery, D. C. (1991). *Design and analysis of experiments* (3rd ed.). New York: Wiley.

Munger, S., Smith, R. W., & Payne, D. (1962). *An index of electronic equipment operability: Data store* (Rep. No. AIR-C43-1/62-RP[1]). Pittsburgh, PA: American Institute for Research.

Newsletter, Test & Evaluation Technical Group, HFES (2002a). Spring, 1–8.

Newsletter, Test & Evaluation Technical Group, HFES (2002b). Autumn, 1–8.

Nielsen, J. (1997). Usability testing. In G. Salvendy (Ed.), *Handbook of human factors and ergonomics* (pp. 1543–1568). New York: Wiley.

Nieva, V. F., Fleishman, E. A., & Rieck, A. (1985). *Team dimensions, their identity, their measurement and their relationships* (Research Note 85-12). Alexandria, VA: Army Research Institute for the Behavioral and Social Sciences.

Park, K. S. (1997). Human error. In G. Salvendy (Ed.), *Handbook of human factors and ergonomics* (pp. 150–173). New York: Wiley.

Pearson, R. G., & Byars, G. E. (1956). *The development and validation of a checklist for measuring subjective fatigue* (Rep. No. TR-56-115). Brooks AFB, TX: School of Aerospace Medicine.

Pejtersen, A. M., & Rasmussen, J. (1997). Effectiveness testing of complex systems. In G. Salvendy (Ed.), *Handbook of human factors and ergonomics* (pp. 1514–1542). New York: Wiley.

Proctor, R. W., & Proctor, J. D. (1997). Sensation and perception. In G. Salvendy (Ed.), *Handbook of human factors and ergonomics* (pp. 43–88). New York: Wiley.

Proctor, R. W., & Van Zandt, T. (1994). *Human factors in simple and complex systems.* Boston: Allyn & Bacon.

Rau, J. G. (1974). *Measures of effectiveness handbook* (Rep. No. AD-A021461). Irvine, CA: Ultrasystems, Inc.

Reason, J. (1987). Generic error-modeling systems (GEMS): A cognitive framework for locating common human error forms. In J. Rasmusson, K. Duncan, & J. Leplat (Eds.), *New technology and human error.* Chichester, England: Wiley.

Richard, G. L., & Parrish, R. V. (1984). Pilot differences and motion cuing effects on simulated helicopter hover. *Human Factors, 26,* 249–256.

Rogers, W. A. (1999). Technology training for older adults. *Proceedings of the CHI'99,* 51–52.

Roscoe, A. H. (1984). Assessing pilot workload in flight. Flight test techniques. *Proceedings of the NATO Advisory Group for Aerospace Research and Development (AGARD)* (Rep. No. AGARD-CP-373). Neuilly-sur-Seine, France.

Rouse, W. B. (1986). On the value of information in system design: A framework for understanding and aiding designers. *Information Processing & Management, 22,* 279–285.

Salvendy, G. (Ed.). (1987, 1997). *Handbook of human factors and ergonomics.* New York: Wiley.

Simon, C. W. (2001). *The Cassandra paradigm: A holistic approach to experiments in the behavioral sciences.* Unpublished paper.

Smode, A. F., Gruber, A., & Ely, J. H. (1962). *The measurement of advanced flight vehicle crew proficiency in synthetic ground environments* (Rep. No. MRL-TDR-62-2). Wright-Patterson AF, OH: Aerospace Medical Research Laboratory.

Stevens, S. S. (1975). *Psychophysics.* New York: Wiley.

Swain, A. D. (1989). *Comparative evaluation of methods for human reliability analysis* (Rep. No. GRS-71). Germany: Gesellschaft fur Reaktor-Sicherheit (GRS) mbh.

Swain, A. D., & Guttman, H. E. (1983). *Handbook of human reliability analysis with emphasis on nuclear power plant applications.* Sandia National Laboratories (NUREG/CR-1278). Washington, DC: Nuclear Regulatory Commission.

Sweetland, J. H. (1988). Beta tests and end-user surveys: Are they valid? *Database, 11,* 27–37.

Thurstone, L. L. (1927). A law of comparative judgment. *Psychological Review, 34,* 273–386.

Van Gigch, J. F. (1974). *Applied general system theory.* New York: Harper & Row.

Van Orden, K. F., Benoit, S. L., & Osga, G. A. (1996). Effects of cold and air stress on the performance of a command control task. *Human Factors, 38,* 130–141.

Virzi, R. A. (1992). Refining the test phase of usability evaluation: How many subjects is enough? *Human Factors, 34,* 457–468.

Weimer, J. (1995). Developing a research project. In J. Weimer (Ed.), *Research techniques in human engineering* (pp. 20–48). Englewood Cliffs, NJ: Prentice-Hall.

Wickens, C. D. (1991). Processing resources and attention. In D. Damos (Ed.), *Multiple task performance* (pp. 1–34). London: Taylor & Francis.

Wickens, D. D. (1992). *Engineering psychology and human performance* (2nd ed.). New York: Harper & Collins.

Wierwille, W. W., & Eggemeier, F. T. (1993). Recommendations for mental workload measurements in a test and evaluation environment. *Human Factors, 35,* 263–281.

Wierwille, W. W., Rahmi, M., & Casali, T. G. (1985). Evaluation of 16 measures of mental workload using a simulated flight task emphasizing mediational activity. *Human Factors, 27,* 489–502.

Williges, R. C. (1995). Review of experimental design. In J. Weimer (Ed.), *Research techniques in human engineering* (pp. 49–71). Englewood Cliffs, NJ: Prentice-Hall.

Winer, B. J., Brown, D. R., & Michels, K. M. (1991). *Statistical principles in experimental design* (3rd ed.). New York: McGraw-Hill.

Woodson, W., Tillman, B., & Tillman, P. (1992). *Human factors design handbook: Information and guidelines for the design of systems, facilities, equipment and products for human use* (2nd ed.). New York: McGraw-Hill.

Zipf, G. K. (1965). *Human behavior and the principle of least effort* (2nd ed.). New York: Hafner.

Author Index

K-L

M

N-O

P-Q-R

S

T-V

W-Z

Subject Index

RECEIVED
NOV 24 2003